11th International Conference on Advances in Mass Data Analysis of Images and Signals in Medicine, Biotechnology, Chemistry and Food Industry (MDA 2016)

New York, New York, USA
July 2016

Editors:

Petra Perner

ISBN: 978-1-5108-9519-5

Printed from e-media with permission by:

Curran Associates, Inc.
57 Morehouse Lane
Red Hook, NY 12571

Some format issues inherent in the e-media version may also appear in this print version.

ibai-publishing
Arno-Nitzsche-Str. 45
04277 Leipzig, Germany

info@ibai-publishing.org

Additional copies of this publication are available from:

Curran Associates, Inc.
57 Morehouse Lane
Red Hook, NY 12571 USA
Phone: 845-758-0400
Fax: 845-758-2633
Email: curran@proceedings.com
Web: www.proceedings.com

Table of Contents

Petra Perner (Ed.)

Advances in Mass Data Analysis of Images and Signals in Medicine, Biotechnology, Chemistry and Food Industry

11th International Conference, MDA 2016
New York, USA, July 2016

Proceedings

ibai Publishing
www.ibai-publishing.org

The German National Library listed this publication in the German National Bibliography.
Detailed bibliographical data can be downloaded from http://dnb.ddb.de.

Printed in Germany, 2016

Preface

The 11th International Conference on Mass Data Analysis of Signals and Images with Applications in Medicine, r/g/b Biotechnology, Food Industries and Dietetics, Biometry and Security, and Agriculture MDA 2016 held in July in New York showed once more that the event is a must for all specialists from research and industry alike who like to stay informed about hot new topics in mass data analysis of signals and images.

We accepted six papers on very important topics. The acceptance rate was 33%. The topics range from Nuclear Medicine, Material Characterization in Production, System Biology, Medical Histological Image Analysis, Video Mining in Agriculture, and Spectrometer-Signal Analysis.

A tutorial on Intelligent Image Interpretation has been held in connection with the conference.
Selected revised papers of this conference as well as contributed papers will be published in the Ocotber issue 2016 of the International Journal Transaction on Mass Data Analysis of Signals and Images.

We are happy to see that new automatic systems for the analysis of signals and images have been presented at MDA 2016 and that many researchers followed our mission to bring new signal and image analysis methods into real applications that introduce a new quality level to various real life applications.

At MDA 2016, one paper become select for Best Paper Award. Congratulations to the nominated authors.

The 12th conference will be held in 2017 in New York (www.mda-signals.de) under the auspices of the World Congress Frontiers on Intelligent Data Analysis DSA 2017 (www.worldcongressdsa.com).
We would like to invite you to contribute to this conference. Please come and join us.

July 2016,

Prof. Dr. Petra Perner
Chair of the Conference MDA

International Conference on Mass Data Analysis with Applications in Medicine, r/g/b Biotechnology, Food Industries and Dietetics, Biometry and Security, and Agriculture MDA 2016

Chair

Committee

Mass Data Acquisition and Analysis of Optical insects' wingbeat signals

Ilyas Potamitis [1*] , Iraklis Rigakis [2] , Spyros Kouzoupis [1]

[1] Dept. of Music Technology & Acoustics, Technological Educational Institute of Crete, E. Daskalaki Perivolia, 74100, Rethymno Crete, Greece
potamitis@staff.teicrete.gr
[2]Dept. of Electronics, Technological Educational Institute of Crete, Romanou 3 Chalepa, Chania, 73133, Greece
rigakis@chania.teicrete.gr
*Author to whom correspondence should be addressed
Tel.: +30 28310 21900; Fax: +30 28310 21914.

Abstract. Flight tones in insects occur during locomotion and courtship and are mostly analyzed using microphones. The use of microphones is impractical for analyzing the wingbeat of non-tethered insects especially if one is interested in studying the frequency content of wingbeats of a large number of insects on the order of hundreds to thousands. In this study we present a practical setting based on a novel 2D optical sensor that we embed inside insectary cages to record the wingbeats of fruit flies and mosquitoes, namely Bactrocera oleae and Aedes albopictus. We show how this setting allows to automatically record large datasets of insect wingbeats that have large economic and social importance. We subsequently show how to automatically analyze such data to create distributions of parameters related to wingbeat frequency and harmonic properties derived from many non-tethered wingbeats and therefore characterize the wingbeat properties of a whole species with increased confidence.

Keywords: wingbeat analysis, precision agriculture

1 Introduction

From an engineering point of view, flying insects can be seen as biological micro-vehicles equipped with multi-sensors able to demonstrate admirable maneuvering capabilities [1]. From another point of view, many species such as fruit flies cause large economic damage to crops. To the top of the hazard list mosquitoes are nuisances and potential carriers of serious vectors for humans and livestock and, therefore, there is a compelling need to develop novel devices and procedures to counter-measure their presence [2]. An informative cue of the presence of these elusive insects is their wingbeat that produces an audible tone. The basic principle of the optical sensor as applied to this particular task is the gradual interruption of the path of light between an emitter and a receiver of light at the rhythm of the wing movement. In [3] we show that though the generative process of audio and optical recordings are completely different, they sound and their associated spectrum looks quite similar. Quality microphone recordings are hard to obtain for insects performing a free flight. Indeed, most reported literature is based on measuring the wingbeat of one or few tethered specimens (see e.g. [4-6]). Tethering is suspected to affect movement as it alters wing kinematics (see [4] and the references to this point therein). Last, microphones inserted in cages pick-up sounds from every direction even if they are directional.

In this work we place a large number of insects of the same species in an insectary along with a novel optoacoustic sensor. The novelty of our sensor with respect to the work presented in [3] is that the receiving part of the sensor is a surface and not a linear array. Due to its optical nature it is not affected by voices and sound interferences commonly encountered in the laboratory and in the field (due to interferences from bird vocalizations, cicadas' songs, mechanical sounds, wind etc). Once enclosing the sensor and the insects in insectary cages insects incidentally and on a random basis cross the sensor and a recording snippet is taken. This settings allows a mass creation of recordings of the order of tens of thousands per day from a single cage. Moreover, it allows us to change temperature and humidity conditions and see what effect they have on the flying activity and properties of the wingbeat.

2 Materials & Methods

2.1 Insects

In this study we include a fast flying insect – a fruit fly and a slower one – the mosquito *Aedes albopictus* also known as 'Tiger mosquito'. Mosquitoes perform erratic movements during flight, they beat their wings at higher rates compared to other insects commonly encountered in domestic environments (e.g. flies, bees, hawkmoths) and are generally slower than them.

B. oleae were collected in Heraklion Crete from olive orchards. The larvae inside the olive fruit are in various instars and they feed upon the pulp until they exit, usually as third instar larvae which pass through the sieve to pupate. Then they are collected and grown in an insectary cage. As larvae turn into adult insects, we supply them with yeast hydrolysate-sugar diet and water to sustain them to life.

Aedes albopictus were collected in Athens (Greece) and maintained in Athens lab for 5 years. Mosquitoes were reared in an environmental chamber at controlled temperature of 27-28 °C, 70-80% humidity and 12h light/dark photoperiod. Cohorts were obtained by hatching eggs in water containers. Larvae were fed with cat food (Purina). Pupae of both sexes were transferred to insect cages with net and offered 20% sucrose solution. Counting from the time of being able to fly *Ae. albopictus* was 3-20 days old.

2.2 Optical sensor

The 2D optoacoustic sensor is shown in Fig. 1. It is composed of a receiver of light placed at the opposite of an infrared light emitter. The emitter is not a single LED but 2x4 infrared ones, operating at 940nm and connected in row. The multiple LEDs solution forming a 2D array were chosen for two reasons: a) to ensure uniform light distribution across the entrance of the trap, b) to increase the field of view of the sensor as a single line of emitters will miss to illuminate insects flying on the border. The receiver is made of 20 photodiodes in a 2x10 arrangement and connected in parallel, thus forming a 2D light receiving surface. Details on the electronics' board used for 1D sensors can be found in [7]. The recordings are normalized in the range −1 and 1 in order to be treated as line-level audio signals (therefore the term opto-acoustic). The sensors carry rechargeable batteries and due to their low power consumption they can record unattended for a number of days, or indefinitely by connecting the kit containing the battery to a charger.

The sensor (see Fig. 2-left) can be used for measuring tethered insects as well but was mainly designed this way in order to be inserted from the entrance window of insectary cages containing flying insects of a single species (see Fig. 2-right). Insects incidentally cross the field of view of the sensor on a random basis and, therefore their wingbeat is recorded. Note that the optical nature of the sensor makes it immune to sound and insects flying outside the field of view. We place several hundred adult insects strictly of a single species in cages and the recording of their wingbeat occurs the moment they pass through the rectangle of the sensors.

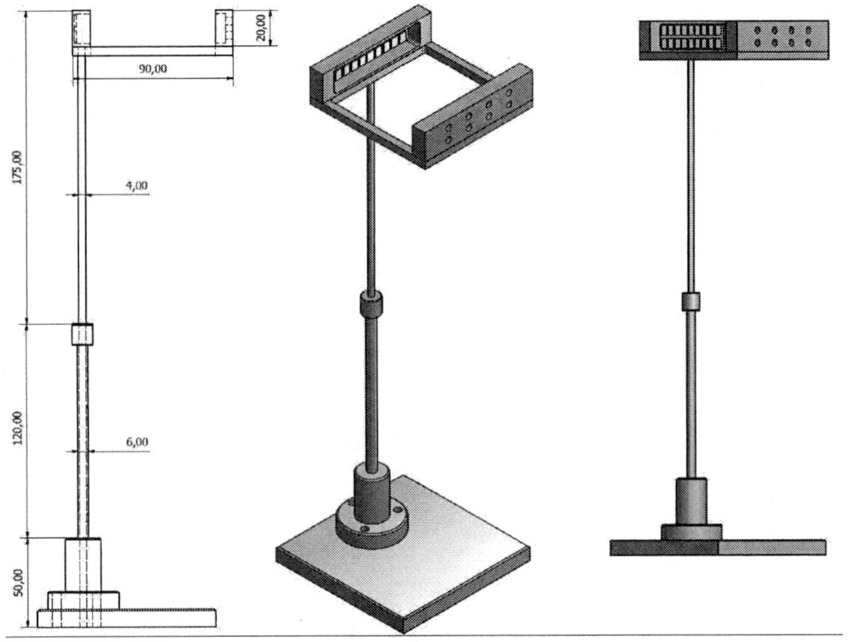

Fig. 1. A schematic diagram of a two-dimensional optoelectronic sensor to record flying insects' wingbeat inside insectary cages. (Left) dimensions, (Middle) a receiving surface of 20 photodiodes placed as two linear layers of 10 diodes each, the emitter: 8 infrared LED, (right) projected view of emitters and receivers.

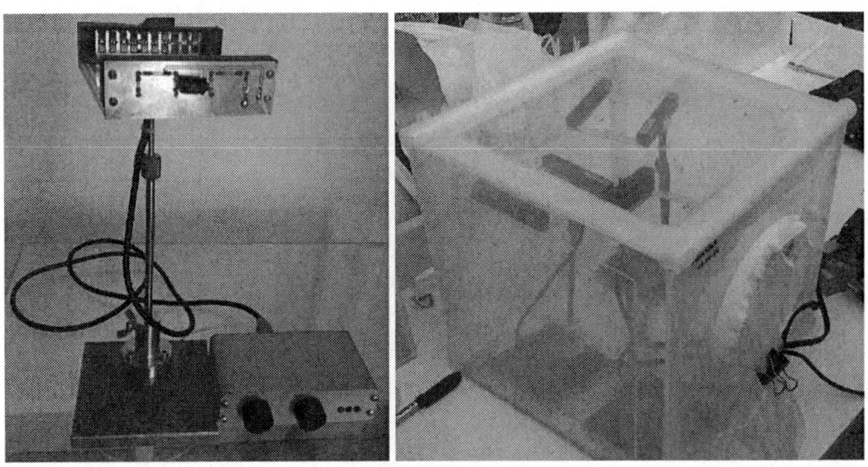

Fig. 2. (Left) A prototype, two-dimensional optoelectronic sensor to record flying insects' wingbeat inside insectary cages. (Right) The sensor in its final standing.

2.3 Analysis of measurements

The power spectral density (PSD) one-sided estimate, of each snippet sampled at 8 kHz is found using Welch's overlapped segment averaging estimator. The signal is divided into sections of length of 256 samples. The modified periodograms are computed using a Hamming window of the same length as the window. The overlapping in windowing equals to the 50% of the window length. The modified periodograms are averaged to obtain the PSD estimate. The FFT length was 256 samples. Fundamental frequency and harmonics were quantified by peak-picking the PSD. The case of *Ae. Albopictus* wingbeat PSD is analyzed in Fig. 3-4.

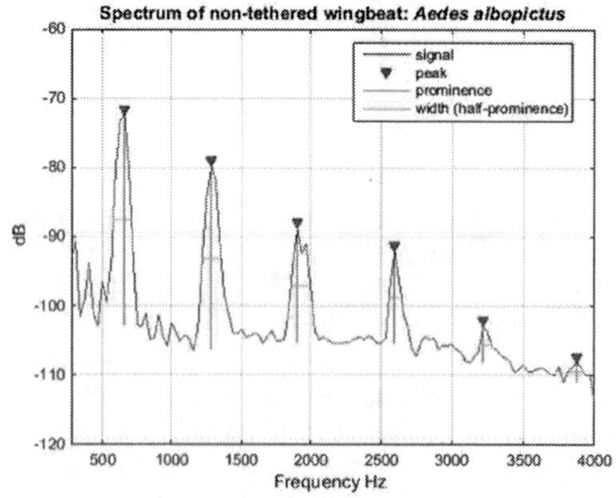

Fig. 3. Spectrum of an optical recording of a single *Ae. albopictus* wingbeat in free flight.

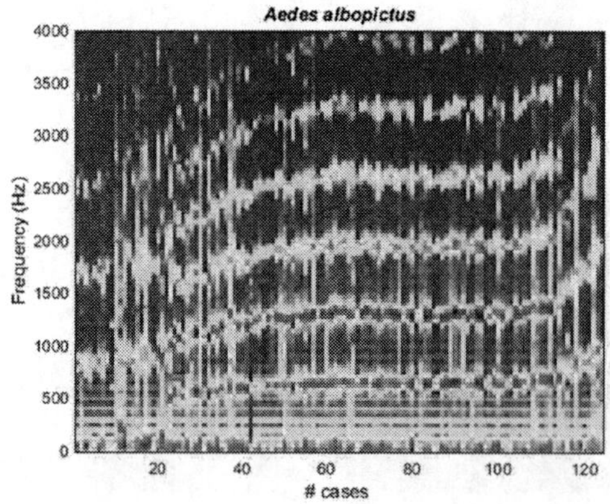

Fig. 4. Spectrum of 124 flight cases of *Ae. albopictus* stacked together and ordered based on the wingbeat frequency from lower to higher values. The first line is the fundamental frequency (500-750 Hz). The other wave patterns are the harmonics. The fundamental and the higher harmonics are clearly resolved.

B. oleae

The case of *B.oleae* wingbeat PSD is analyzed in Fig. 5 and Fig. 6. In Fig. 5 we show a snippet in the time domain corresponding to a wingbeat recording of a flight case and in Fig. 6 we stack together 120 recordings of wingbeats of *B. oleae*.

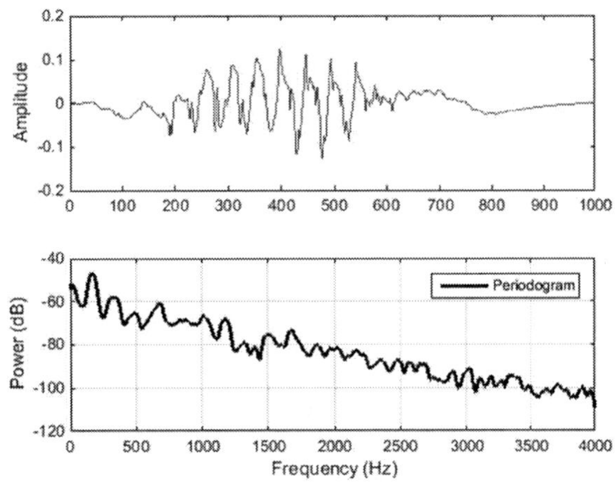

Fig. 5. (Top) Plot of a wingbeat snippet of B. oleae. (Bottom) Spectrum of an optical recording of a single *B. oleae* in free flight.

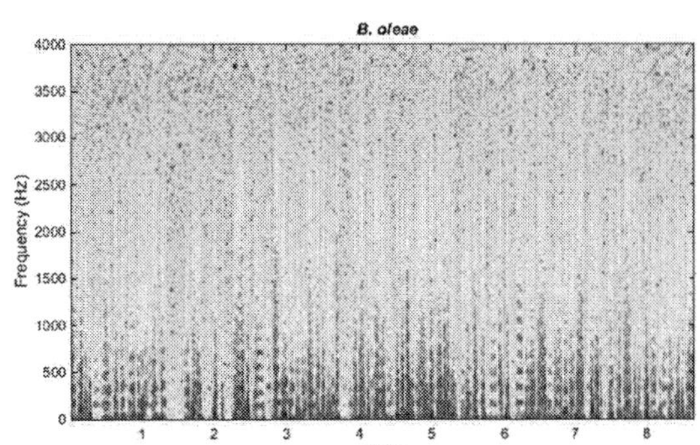

Fig. 6. Spectrum of 120 flight cases of *B. oleae* stacked together. One can see the fundamental around 180-200 Hz and the overtones. Frequencies lower than 100 Hz are due to body movement.

3 Discussion

In this study we show a practical setting to record free flight of insects of economic and social importance using optoacoustic sensors. This choice of these sensors and the fact that can be inserted in insectary cages permits effortless acquisition of massive datasets of wingbeat recordings of insects in flight and allows the automatic analysis of their spectral content as it returns discrete snippets of each recording.

4 Conclusions

We have shown a novel way to create massive datasets of wing beat signal of any insect that can be in flight. The practical applications of this technology include optical detectors that embedded in traps report counts and species composition of trapped adults and feed wirelessly these data to epidemiological models that monitor the current situation but also predict future outbreaks based on past count and species distribution.

5 Acknowledgments

The research leading to these results has received funding from the European Union's HORIZON 2020 fast track to innovation project under grant agreement n°691131 project REMOSIS.

References

1. Ellington C., (1999). The novel aerodynamics of insect flight: applications to micro-air vehicles. Journal of Experimental Biology, Dec; 202(Pt 23):3439-48.
2. Biswas, A.; Siddique, N.; Bing Bing Tian; Wong, E.; Caillouet, K.; Motai, Y. Design of a Fiber-Optic Sensing Mosquito Trap, in Sensors Journal, IEEE, vol. 13, no.11, pp. 4423-4431, 2013. doi: 10.1109/JSEN.2013.227427
3. Potamitis, I. Rigakis, I. Novel Noise-Robust Optoacoustic Sensors to Identify Insects Through Wingbeats. IEEE Sensors Journal, vol.15, no.8, pp. 4621-4631, Aug. 2015, doi: 10.1109/JSEN.2015.2424924

4. Arthur B, Emr K, Wyttenbach R, Hoy R. Mosquito (Aedes aegypti) flight tones: Frequency, harmonicity, spherical spreading, and phase relationships. The Journal of the Acoustical Society of America. 2014;135(2):933-941. doi:10.1121/1.4861233.
5. Aldersley A., Champneys A., Homer M., Robert D., (2014). Time-frequency composition of mosquito flight tones obtained using Hilbert spectral analysis, The Journal of the Acoustical Society of America, vol. 136, issue 4, pp. 1982-1989, 10.1121/1.4895689
6. Perevozkin V., Bondarchuk S., Species specificity of acoustic signals of malarial mosquitoes of Anopheles maculipeniis complex, International Journal of Mosquito Research 2015, 2(3), pp. 150-155.
7. Potamitis I, Rigakis I, Fysarakis K (2015) Insect Biometrics: Optoacoustic Signal Processing and Its Applications to Remote Monitoring of McPhail Type Traps. PLoS ONE 10(11): e0140474. doi:10.1371/journal.pone.0140474

Segmentation Approaches for Human Metaspread Chromosome Images Using Level Set Methods

Tanvi Arora[1], Dr. Renu Dhir[2]

[1] Research Scholar,Dr. B.R Ambedkar National Institute of Technology, Jalandhar, Punjab ,India

[2] Associate Professor, Dr. B.R Ambedkar National Institute of Technology, Jalandhar, Punjab, India

tanviverma@rediffmail.com1,dhirr@nitj.ac.in

Abstract. The human metaspread images contain the chromosomes of an individual that are imaged during the metaphase stage of cell division. The chromosomes are the genetic information carriers, any alteration either in the number of chromosomes or the structure of chromosomes, results in a medical condition termed as genetic defects. The genetic defects can be cause of many diseases that are difficult to cure. In order to uncover the genetic defects the metaspread images are segmented, to count the number of chromosomes present or to study the structure of the chromosomes. The metaspread images suffer from intensity in homogeneity, because of which it is very difficult to extract the individual chromosomes by using conventional segmentation techniques. In this work various segmentation approaches are implemented and studied, to segment the chromosomes from the metaspread images. The minimization of region scalable fitting energy for image segmentation proposed by Chumming li et. al, is effective in segmenting the metaspread images that have intensity in homogeneity, as this technique uses the local intensity values of the nearby regions of the objects and find the approximate intensity values along both sides of the contour. The paper compares the level set based segmentation algorithms based upon their implementation techniques and the segmentation results. The methods have been compared in terms of number of objects segmented, time taken, complexity, type of segmentation approach. The ADIR dataset of metspread images has been taken for the purpose of experimentation.

Keywords: Metaspreads, Chromosomes, Segmentation, Level Set Methods.

1 Introduction

The human metaspread images contain the chromosomes that are imaged using a light microscope[29] during metaphase of mitosis. The cells are usually imaged during the metaphase or the pro metaphase as during this phase only the chromosomes are visible as distinct bodies and can give useful information[18].The cells that are imaged for the metaspread image generation are generally taken from the samples of blood, product of conception, amniotic fluid, tissues or bone marrow[6][19][32]. A normal healthy human being has 23 pairs of chromosomes out of which 22 pairs are homologous and the 23rd pair has sex determining chromosomes that are either XX or XY for females and males respectively[20].Any alteration either in structure or the number of the chromosomes is termed as genetic defect[14]. The genetic defects are the underlying cause of mental retardation, frequent miscarriage, epileptic seizures or various forms of cancers[15].

In order to extract the chromosomes from the metaspread images, the human metaspread images undergo the segmentation process. The human metaspread images suffer from intensity in homogeneity[16] due to which the conventional segmentation approach based upon global thresholding[31] are not suitable for metaspread image segmentation, as in case of very small chromosomes or towards the tips of the chromosomes the intensity is very low. The global thresholding based methods can miss these portions. Secondly the chromosomes are non-rigid objects and are generally present in various orientations[33], they may be bent, touching each other or may be overlapping[5], global thresholding based method are not able to segment these types of chromosomes and thus can segment them as a cluster of chromosomes as evident from the fig 1 below.

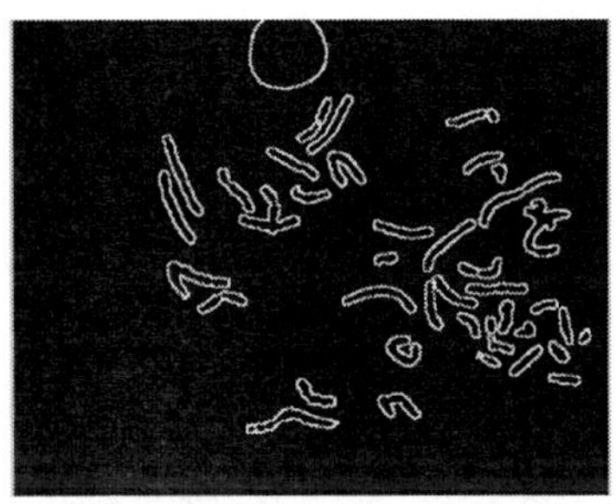

Fig. 1. Metaspread Image as segmented by Otsu method

From last so many years many methods have been proposed for the segmentation of chromosome metaspread images. Most of the methods have used Otsu thresholding based methods for the purpose of segmentation[4][41][21][43]. The Otsu segmentation approach is based upon the global thresholding approach of segmentation, this technique is not well suited for segmentation of metaspread images, as metaspread images have intensity in homogeneity and thus results in either over segmentation or under segmentation.

In order to deal with the drawbacks of global thresholding, a local thresholding based approach termed as adaptive thresholding was proposed[44]. This technique is quite successful in segmenting the images that have intensity inhomogeneity[34][9][28][17] but it fails in case of chromosomes that are lying close to each other, it segments them as a cluster of chromosomes. In a way it under segments the metaspread images for the objects that are lying close to each other. The segmentation results for metaspread images for adaptive thresholding are illustrated in fig 2 below

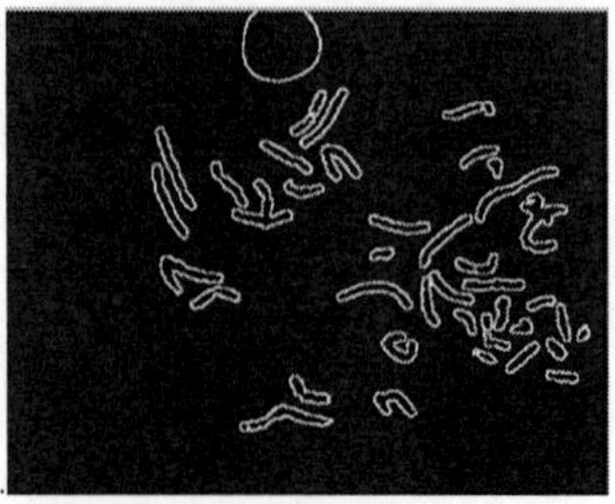

Fig. 2. Segmentation of metaspread images using adaptive thresholding

Watershed transform based segmentation techniques have also been used for segmenting the metaspread images[22][23][40][3][2] these techniques find the distance transform of the image to determine the number of regions to be segmented, this results in over segmentation due to local minima's, so in order to avoid local minima's before applying the watershed segmentation the negative distance transformed image is reconstructed. The watershed based segmentation technique is computationally expensive; it results in under segmentation of the metaspread images.

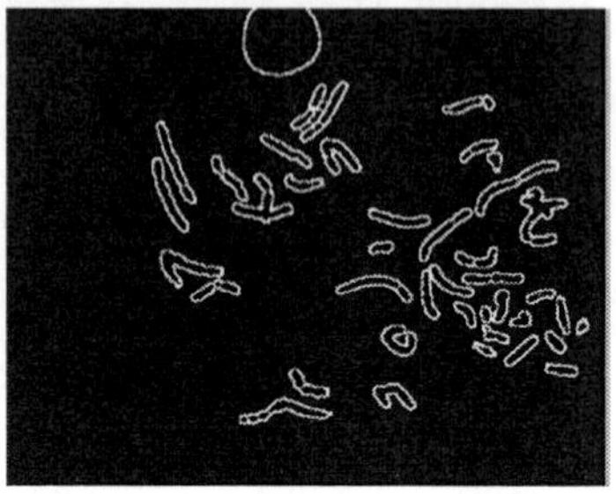

Fig. 3. Results of segmentation after application of watershed method

Most of the previously reported methods have either used Otsu method, adaptive thresholding or watershed methods for the segmentation of the chromosomes, and much effort has been done in reconstructing the overlapping and touching chromosomes as reported in[38][37][39][11][35]. Since the metaspread images as shown in fig 3 below, contains intensity in homogeneities, the level set based segmentation techniques have proved to be quite fruitful to segment the metaspread chromosome images as reported in[34]. These techniques segment the images by drawing a smooth contour around the boundaries of the object and are very efficient in identifying the objects present in the image irrespective of the shape and size. Over the years many level set based methods have been proposed for image segmentation. In this work an effort has been made to compare and contrast various level set based segmentation algorithms, and propose the most suitable algorithm for the segmentation of metaspread images.

Im-age						
Im-age nam e	A0102XYD	A020315D	P0802XYD	P110121D	P110536D	P1102XYD

Fig. 4. Sample metaspread images depicting intensity inhomogeneity

The rest of the paper is organized as follows: section 2 explains material and methods, section 3 has results and discussion and section 4 concludes the paper.

2 Material and Methods

2.1 Dataset Used

In order to evaluate the various segmentation algorithms the DAPI images of the ADIR dataset[46]has been used. The dataset has 200 images. For the purpose of our study we have taken all the images. The result that are presented in this paper are for a single image named as P110536D as shown in fig 3 above.

2.2 Performance Evaluation

The performance of segmentation can be evaluated by comparing the segmentation results with the ground truth information. The measure for performance evaluation is accuracy, which can be calculated as follows:

$$Accuracy = \frac{Count\ of\ correctly\ segmented\ individual\ \ objects}{Total\ Number\ of\ individual\ Objects} \tag{1}$$

The results of the segmentation have been compared with the ground truth information provided by experienced cytogeneticists.

2.3 Methods

This section describes the various segmentation techniques that have been taken for the purpose of segmentation of the metaspread images. The proposed techniques have been implemented using the MATLAB 2014 as per their original manuscript details. The software named as Creatis [7] has been used for the evaluation of some of the methods.

2.4 Level Set Based Segmentation

Level set methods are a kind of deformable models for segmentation of images, in which the initial level set is initialized, then based upon the properties of the image and the level set, the level set is recomputed, and is moved towards the boundaries of the objects of the image[13][42]. There are generally two type of properties that helps in the evolution of the level set, one is based upon the internal properties defined within the level set that helps the boundaries of the curve to be smooth and the second are the external properties that are dependent upon the image, that helps the curve to be more towards the boundaries of the objects in the image[42]. They are based upon implicit deformable models and are capable of handling the topological variations[3]. Over the years many level set based algorithms have evolved for the segmentation of images that suffer from intensity inhomogeneity[25]. In this paper 10 level set based methods have been compared.

2.3.1 Geodesic Active Contours[10]

This method is based upon image gradient, which is used to calculate the force function. The curve is converged towards regions of higher gradient value. It has a inbuilt regularization term, the curve evolution equation is calculated using distance function, every iteration re initialise the evolution equation.
The energy for the level set is calculated using the following equation

$$E_t(\Gamma) = \int_0^1 g_t\left(Im(\Gamma(q))\right) \|\Gamma'(q)\| dq \quad (2)$$

Where

$$g_t(Im) = \frac{1}{1+\|\nabla(G*Im)\|^2} \quad (3)$$

$Im(.)$ Is the intensity of the image I, Γ represents the parametric curve, and G is the filter of the type Gaussian with unit variance.
The evolution equation for this method is

$$\frac{\partial\phi}{\partial t}(x) = g_t(Im(x))\|\nabla\phi(x)\|(c+\kappa) + \nabla g_t(Im(x))\nabla\phi(x) \quad (4)$$

Where $\kappa = div\left(\frac{\nabla\phi(x)}{\|\nabla\phi(x)\|}\right)$ represents the curvature of the emerging curve and c is the balloon force and a constant value that is set to 1.

2.3.2 Active Contours without Edges[12]

This approach aims at dividing the image into two regions that are homogeneous based upon the average value. This method is sensitive to initialization as the evolution equation is computed on level sets, one of the narrow bands. The curve evolution equation is re computed for each iteration.
The energy for the level set is calculated using the following equation:

$$\mathrm{E_t}(\phi) = \int_\Omega F_t\left(Im(x), \phi(x)\right) dx + \lambda \int_\Omega \delta\left(\phi(x)\right) \|\nabla\phi(x)\| \, dx \quad (5)$$

Where δ represents the dirac function and

$$F_t\left(Im(x), \phi(x)\right) dx = H_t\left(\phi(x)\right)(Im(x) - \upsilon)^2 + \left(1 - H_t\left(\phi(x)\right)\right)\left(Im(x-u)\right)^2 \quad (6)$$

Where H is Heaviside function, v *and* u are the parameters to be updated on every iteration

$$u = \frac{\int_\Omega \left(1 - H_t(\phi(x))\right).Im(x)\,dx}{\int_\Omega 1 - H_t(\phi(x))\,dx} \quad (7)$$

$$v\frac{\int_\Omega \left(H_t(\phi(x))\right).Im(x)\,dx}{\int_\Omega H_t(\phi(x))\,dx} \quad (8)$$

$E(\phi)$is a combination of data attachment term and regularization term that aims at minimizing the length of the contour and smoothing it during its evolution. The evolution equation for this method is as follows:

$$\frac{d\phi}{dt} = \delta(\phi(x))((Im(x) - v)^2 - (Im(x) - u)^2) + \lambda\delta(\phi(x))\kappa \ (9)$$

2.3.3 Minimization of Region Scalable Fitting Energy for Image Segmentation[26]

The proposed technique overcomes the segmentation difficulties that arise due to intensity inhomogeneity. It is based upon the region based active contours, in which the local intensity information is retrieved at a controllable scale. The energy criterion is defined in form of a contour and in addition to that two local energy fitting functions are used to find the local intensities along both sides of the contour. A level set formulation is created with level set regularization term using which the equation of the curve is evolved for minimization of the energy. The extraction of the local fitting energy helps in dealing with the intensity in homogeneity issue. There is no re-initialization of the level set function as it is preserved by the regularization term. The evolution equation is based upon the signed distance transform.
The energy criterion is based upon the following equation:

$$\begin{aligned} E_t(\phi) = \lambda_1 \int_\Omega \int_\Omega K_{t_\sigma}(x \\ - y)|Im(y) - f_1(x)|^2 H_t(\phi(x))\,dy\,dx + \lambda_2 \int_\Omega \int_\Omega K_{t_\sigma}(x \\ - y)|Im(y) - f_2(x)|^2 (1 - H_t(\phi(x)))\,dy\,dx \\ + v \int_\Omega \delta(\phi(x))\|\nabla\phi(x)\|\,dx \\ + \mu \int_\Omega \frac{1}{2}(\|\nabla\phi(x)\| - 1)^2\,dx \quad (10) \end{aligned}$$

Where $Im(x)$is the intensity of the image, H is the Heaviside function, K_σ is the Gaussian kernel.

$$K_{t_\sigma}(u) = \frac{1}{(2\pi)^{n/2}\ \sigma^n} e^{-\|u\|^2/2\sigma^2} \quad (11)$$

Where σ is a scale parameter greater than 0 and f_1,f_2 are the functions for the pixel located at x and calculated for each iteration as:

$$f_1 = \frac{K_{t_\sigma} * \left(H_t(\phi(x))Im(x)\right)}{K_{t_\sigma} * H_t(\phi(x))} \quad (12)$$

$$f_2 = \frac{K_{t_\sigma} * \left(1 - H_t(\phi(x))Im(x)\right)}{K_{t_\sigma} * \left(1 - H_t(\phi(x))\right)} \quad (13)$$

λ_1 and λ_2 are the constants. The energy term has four integrals, the first and second represents the data terms; the third is regularization term used to smoothen the curve and fourth is the regularization term that forces the level set to maintain the signed distance properties during the evolution.
The evolution equation used for this method is :

$$\frac{\partial\phi}{\partial t} = \delta(\phi(x))\left(\lambda_1 \int_\Omega K_{t_\sigma}(x-y)|Im(y) - f_1(x)|^2\, dy + \lambda_2 \int_\Omega K_{t_\sigma}(x-y)|Im(y) - f_2(x)|^2\, dy\right) + \upsilon\delta(\phi(x))\kappa + \mu(\nabla^2\phi(x) - \kappa) \quad (14)$$

2.3.4 Localizing Region Based Active Contours[24]

It is a region based segmentation approach, in which local image information is considered and the contour is evolved using the local information. The heterogeneous image can be segmented due to localized contours. This method is sensitive to initialization. The evolution function is based upon distance transform and is recomputed for every iteration.
The energy criteria for this method are:

$$E_t(\phi) = \int_\Omega \delta(\phi(x)) \int_\Omega B_t(x,y) \cdot F_t(Im(y), \phi(y))\, dy\, dx + \lambda \int_\Omega \delta(\phi(x))\, \|\nabla\phi(x)\|\, dx \quad (15)$$

Where Dirac function is denoted by δ and B represents the Ball function with radius r for the point x, it is defined as:

$$B_t(x,y) = \begin{cases} 1, \|x-y\| \le r \\ 0, otherwise \end{cases} \quad (16)$$

and

$$F_t(Im(y), \phi(y)) = \begin{cases} H_t(\phi(y))(Im(y) - v(x))^2 + \left(1 - H_t(\phi(y))\right)(Im(y) - u(x))^2, ChanVesefeatures \\ (v(x) - u(x))^2 \quad Yezzifeatures \end{cases} \quad (17)$$

Where H represents the Heaviside function, $v(x)$ and $u(x)$are the functions that are to be updated at each iteration and are defined as follows:

$$u(x) = \frac{\int_\Omega B_t(x,y) \cdot \left(1 - H_t(\phi(y))\right) \cdot Im(y)\, dy}{\int_\Omega B_t(x,y) \cdot \left(1 - H_t(\phi(y))\right) dy} \quad (18)$$

$$v(x) = \frac{\int_{\Omega} B_t(x,y) \cdot H_t(\phi(y)) \cdot Im(y)\, dy}{\int_{\Omega} B_t(x,y) \cdot H_t(\phi(y))\, dy} \quad (19)$$

The energy term is the combination of data term and the regularization term.
The evolution equation for the method is given by the equation as below:

$$\frac{\partial \phi}{\partial t}(x) = \delta(\phi(x)) \int_{\Omega} B_t(x,y) \cdot \nabla_{\phi} F_t(I(y), \phi(y))\, dy + \lambda(\delta(x))\kappa \quad (20)$$

Where

$$\nabla_{\phi} F_t(I(y), \phi(y)) = \begin{cases} \delta(\phi(y)) \left((I(y) - v(x))^2 - (I(y) - u(x))^2 \right) & \text{Chan } Vese feature, \\ \delta(\phi(y)) \left(\frac{(I(y) - v(x))^2}{A_v} - \frac{(I(y) - u(x))^2}{A_u} \right) & \text{Yezzi feature,} \end{cases} \quad (21)$$

A_v and $A_u$represent the local exterior and interior areas as calculated by:

$$A_u = \int_{\Omega} B_t(x,y) \cdot \left(1 - H_t(\phi(x))\right) dy \quad (22)$$

$$A_v = \int_{\Omega} B_t(x,y) \cdot \left(H_t(\phi(x))\right) dy \quad (23)$$

2.3.5 A Real Time Algorithm for the Approximation of Level Set Based Curve Evolution[36]

This is a very fast approximation algorithm based upon curve evolution for level sets. It is a two-step method and is independent of partial differential equations. First step aims at evolution of the curve for the data dependent parameter and the second step smoothens the regularization term. The evolution equation for the proposed method can be represented as below:

$$F_t(Im(x), \phi(x)) = H_t(\phi(x))(Im(x) - v)^2 + \left(1 - H_t(\phi(x))\right)(Im(x) - u)^2 \quad (24)$$

Where H is Heaviside function, u and v represents the parameters that are recomputed for each iteration as per equations of Chan Vese model.

2.3.6 Variational B-Spline Level Set[8]

This is a region based segmentation method that aims at dividing the input image into two classes of homogeneous regions. In this the energy computation function is a continuous parametric based upon B-Splines. The minimization term is computed with the coefficients of the B Spline.
The energy criterion for this model is given by the equation below:

$$E_t(\phi) = \int_{\Omega} F_t(Im(x), \phi(x))\, dx, \quad (25)$$

The data attachment term is given by

$$F_t(Im(x), \phi(x)) = H_t(\phi(x))(Im(x) - v)^2 + \left(1 - H_t(\phi(x))\right)(Im(x) - u)^2 \quad (26)$$

Where H is Heaviside function, u and v represents the parameters that are recomputed for` each iteration as per equations of Chan Vese model.

The minimization of the energy function can be carried with respect to the coefficients of B-spline c[k]. The derivatives can be expressed using the equation given as under:

$$\frac{\partial E}{\partial c[K_{t_o}]} = \int_{\Omega} \frac{\partial F_t(x, \phi(x))}{\partial \phi(x)} \cdot \beta^n \left(\frac{x}{h} - K_{t_0}\right) dx \qquad (27)$$

With

$$\frac{\partial F_t(x, \phi(x))}{\partial \phi(x)} = \delta(\phi(x))((Im(x) - v)^2 - (Im(x) - u)^2) \quad (28)$$

The level set evolution can be calculated using the gradient descent on the coefficients of B-spline. The coefficients of B-spline are given as:

$$c^{i+1} = c^i - \lambda \nabla_c E_t(c^i) \quad (29)$$

λ represents the step of the iteration and ∇_c denotes the gradient energy.

2.3.7 Active Contours with Selective Local or Global Segmentation: A New Formulation and Level Set Method[45]

It is a region based active contour technique for image segmentation. It uses a binary level set function and a Gaussian based regularization term. It uses a region based function which stops the contours at weak edges. It can perform local as well as global segmentation.

The signed pressure force (SPF) function is calculated using the equation below:

$$spf_t(Im(x)) = \frac{Im(x) - \frac{c_1 + c_2}{2}}{\max\left(\left|Im(x) - \frac{c_1 + c_2}{2}\right|\right)} \quad x\epsilon\Omega, \qquad (30)$$

Where c_1 and c_2 are calculated as

$$c_1(\phi) = \frac{\int_{\Omega} Im(x) \cdot H_t(\phi)\, dx}{\int_{\Omega} H_t(\phi)\, dx} \qquad (31)$$

$$c_2(\phi) = \frac{\int_{\Omega} Im(x) \cdot (1 - H_t(\phi)\, dx)}{(1 - H_t(\phi)\, dx)} \qquad (32)$$

The level set formulation can be done using the equation below:

$$\frac{\partial \phi}{\partial t} = spf_t(Im(x)) \cdot \alpha|\nabla \phi|, x\varepsilon\Omega \quad (33)$$

The regularization of the level set is done using the Gaussian filter (G_σ) as

$$\phi = \phi * G_{t_\sigma} \qquad (34)$$

2.3.8 Medical Image Segmentation Based on a Hybrid Region Based Active Contour Model[27]

This is a hybrid algorithm based upon the local and global parameters, in order to deal with the issues of intensity in homogeneity. The energy function used in this approach is a weighted sum of local, global and regularization parameters. The curve evolution equation is derived by minimizing the energy.
The energy function is a hybrid energy function computed by:

$$E_t^{Hybrid} = \alpha E_t^{Global} + \beta E_t^{Local} + \omega L_t \qquad (35)$$

Where $L(\phi)$ is the regularization term.

$$E_t^{Hybrid}(\phi, m, n, u_x, v_x) = \alpha \int_{\Omega_y} \left(H_t\phi(y)(Im(y) - m)^2 + (1 - H_t\phi(y))(Im(y) - n)^2\right) dx\, dy + \beta \int_{\Omega_x} \int_{\Omega_y} B_t(x,y)\,(u_x - v_x)^2\, dx\, dy + \omega \int_{\Omega_x} \delta\phi(x)|\nabla\phi(x)|\, dx \qquad (36)$$

The level set function can be minimized by using the equation as:

$$\frac{\partial\phi}{\partial t}(x) = \alpha\delta\phi(y)[-(m-n)(2Im - m - n)] + \beta \int_{\Omega_x} B_t(x,y)\delta\phi(y) \times \left(\frac{(Im(x) - u_x)^2}{A_u} - \frac{(Im(x) - v_x)^2}{A_v}\right) dy + \omega\delta\phi(x) div\left(\frac{\nabla\phi(x)}{|\nabla\phi(x)|}\right) \qquad (37)$$

A_u represents the local interior area and $A_v$represents the local exterior area.

2.3.9 Region Based Segmentation in Presence of Intensity in Homogeneity Using Legendre Polynomials[30]

This approach of segmentation is based upon the Chan-Vese approach in which the energy function is represented as:

$$\varepsilon_t(\phi, c_1, c_2) = \int_\Omega |f_t(x) - c_1|^2 H_{t_\varepsilon}(\phi)\, dx + \int_\Omega |f_t(x) - c_2|^2 \left(1 - H_{t_\epsilon}\phi(x)\right) dx \qquad (38)$$

It improves the energy function $\varepsilon_t(\phi, c_1, c_2)$by replacing the constants $c_1$and c_2 by two smooth functions $c_1^m(x)$and $c_2^m(x)$. These smooth functions are represented as a combination of some Legendre functions.
The energy function is given by

$$\varepsilon_t{}^m(\phi, A, B) = \int_\Omega |f_t(x) - A^T \mathbb{P}_t(x)|^2 \, m_1(x) \, dx + \lambda_1 \|A\|_2^2$$
$$+ \int_\Omega |f_t(x) - B^T \mathbb{P}_t(x)|^2 \, m_2(x) \, dx + \lambda_2 \|B\|_2^2$$
$$+ v \int_\Omega \delta(\phi) \frac{\nabla\phi}{|\nabla\phi|} dx \qquad (39)$$

The curve evolution is performed with the following equation:

$$\frac{\partial\phi}{\partial t} = [-|f_t(x) - A^T \mathbb{P}_t(x)|^2 + |f_t(x) - B^T \mathbb{P}_t(x)|^2]\delta_\varepsilon(\phi)$$
$$+ v\delta_\varepsilon(\phi) div\left(\frac{\nabla\phi}{|\nabla\phi|}\right) \qquad (40)$$

2.3.10 Segmentation of Regions of Interest Using Active Contours with SPF Function [1]

The proposed method aims at segmenting the image into any number of sub regions, and then step by step each region is considered. A signed pressure force (SPF) function is used in the level set formulation. The procedure keeps on dividing the sub regions iteratively and discards the outer regions till some stopping criteria is met. It does not require re-initialization.

The energy function is calculated by using the equation as given below:

$$E_{t_{proposed}} = \int_\Omega |Im(x) - c_1|^2 \, H_{t_\varepsilon}(\phi(x)) M_t{}^k(x) \, dx$$
$$+ \int_\Omega |Im(x) - c_2|^2 \left(1 - H_{t_\varepsilon}(\phi(x))\right) M_t{}^k(x) \, dx \quad (41)$$

The level set is regularized using the Gaussian kernel as

$$\phi^n = G_{t_\sigma} * \phi^n \qquad (42)$$

3 Results and Discussion

For the purpose of comparison a sample image from the ADIR dataset has been taken as shown below in fig 5 below. The ground truth for the image taken is as given in table 1. As per the ADIR dataset the image named P110536D, has 43 chromosomes and 2 noisy objects as depicted in figure. Out of 43 chromosomes 39 chromosomes are individual and 4 chromosomes are present as two clusters having 2 chromosomes each.

The sample image was segmented using the 8 level set based algorithms that has evolved over the years and the results of comparison based upon the segmentation results are shown in the table 2 below. The segmentation results have been compared for the parameters of total number of iterations, total number of objects segmented, count of individual chromosomes segmented, count of noisy objects segmented, count of cluster of chromosomes segmented, and finally accuracy is measured in terms of correctness of number of individual chromosomes segmented.

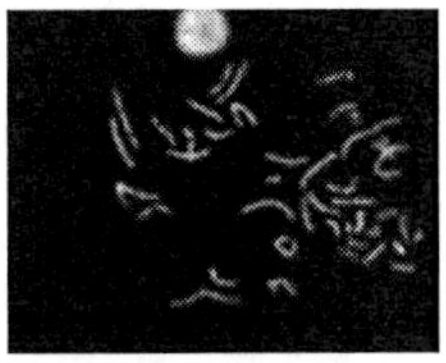

Fig. 5. Image P110536D from ADIR dataset

Object Present	45
Noisy objects	2
Total Chromosomes	43
Individual Chromosomes	39
Cluster of Chromosomes	2

Table 1 Ground truth for Sample Image

Seeing the results of table 2 below the fastest method for level set segmentation is Variational B-spline level-set: a linear filtering approach for fast deformable model evolution.[8], the methods proposed by [26]&[36] are also quite fast as compared to others. But while implementing [36] in MATLAB environment it was quite slow, but is worked quite well in C++ environment. The segmentation results for the segmentation of the metaspread images were best with the [26], it was capable of segmenting the metaspread image with 100% accuracy, but this method has a limitation that it results in over segmentation in case of noisy object.

The method proposed by [10] is very slow and do not work well in case of objects which are lying close by, it segments them as a cluster of chromosomes. In addition to that the accuracy of segmentation is very low.

Most of the proposed segmentation methods based upon level set are enhancements of the [12]. It is quite good at the segmentation, is able to segment the nearby objects with good accuracy. The major drawback of this method is the speed. The segmentation accuracy is at par with the recently proposed methods. The method proposed by [24] is based upon [12], but was not able to segment the metaspread images, it is suitable for segmenting a single object only. It is slow and is heavily dependent upon the initial contour.

As per the literature survey the fastest of all the methods till the year 2008 was [36],as it is based upon the linked list data structure, but it was not able to work well while segmenting the metaspread chromosome images. It is very much dependent upon the initial contour. It is not able to segment the objects that are lying close to each other.

A B-Spline based very fast method has been proposed by [8], it is heavily dependent upon initial contour. It was not able to segment the objects lying close to each other. It is fastest of all the methods, but is not suitable for the segmentation of metaspread chromosome images.

Recently a method has been proposed by [30], it is uses Legendre polynomials, but this method is very slow. But it is able to segment the objects with good accuracy.

The method proposed by[45] based upon local and global segmentation is fast. Is not dependent upon the initial contour value and can segment the chromosomes with quite good accuracy. It is the near competitor of the best method from the selected methods in terms of speed and accuracy of segmentation. Form the comparison of table 2 below the best performer is the [26], which is able to segment the metaspread chromosomes with 100% accuracy. The only drawback of this method is that it results in over segmentation in case of very large noisy objects, as these large noisy objects have varying intensity values due to which this over segmentation results.

Table 2. Results of segmentation on application of chosen level set based algorithms

S No	Parameter	[10]	[12]	[26]	[24]	[36]	[8]	[45]	[30]
1	Segmentation results								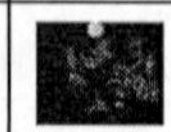
2	Speed	Slow	Slow	Fast	Slow	Very Fast	Very Fast	Fast	Slow
3	Number of Iterations	4000	3810	186	561	34	10	220	2178
4	Count of Objects segmented	23	41	47	1	23	33	41	40
5	Count of Noisy objects	3	2	6	1	1	1	2	1
6	Count of Individual Chromosomes	13	36	39	0	21	25	36	36
7	Count of Cluster of Chromosomes	7	3	2	0	1	7	3	3
8	Accuracy	33.3%	92.3%	100%	0%	53.8%	64.1%	92.3%	92.3%

In the table 3 below the various level set based methods have been compared based upon the underlying implementation details. The parameters used for the comparison of the various models are :*Based upon, Energy , Speed, Evolution, Threshold Energy, Re-initialization, PDE, Energy Terms ,initialization, Regularization Term.*

Table 3 Comparative analysis of the chosen algorithms

S No	Parameter	[10]	[12]	[26]	[24]	[36]	[8]	[45]	[27]	[30]	[1]
1	Based upon	Geodesics curves	Mumford shah model	Gaussian Geodesic curves	Chan Vese &Yezzi	Approximation of level set	B-Splines & Mumford Shah model	Geodesics curves and Chan Vese	Chan Vese , Linkton& Yezzi	Chan Vese	Chan Vese & C Li
2	Energy	Contour Based	Region Based	Localized region Based	Localized region based	Region Based	Region Based	Localized Region Based	Region Based	Localized Region Based	Localized Region Based
3	Speed	Slow	Slow	Fast	Slow	Fast	Fast	Fast	Fast	Slow	Fast

4	Evo-lution	Nar-row Band	Nar-row Band	Whole Do-main	Nar-row Band	Nar-row Band	Whole Do-main	Local & Global, its op-tional	Nar-row Band	Nar-row Band	Nar-row band
5	Threshold En-ergy	Local	Global	Local	Local	Local	Local	Local	Hy-brid	Local	Local
6	Re-ini-tiali-zation	Yes	Op-tional	No	Yes	No	No	No	No	No	Yes
7	PDE	Yes	Yes	No	Yes	No	No	Yes	No	Yes	Yes
8	En-ergy Terms	1	2	3	2	2	2	2	3	3	2
9	ini-tiali-zation	Yes	No	No	Yes	Yes	No	No	Yes	Yes	Yes
10	Regu-lari-zation Term	Yes	Yes	Yes	Yes	Yes	No	Yes	Yes	Yes	Yes

4. Conclusions

In this work various level set based segmentation algorithms have been examined for their suitability for the segmentation of the metaspread images of the human chromosomes. The metaspread images suffer from intensity inhomogeneity and are therefore difficult to segment with the normal segmentation methods. The level set based methods are quite efficient at segmentation of the images with intensity in homogeneity. As they draw the intensity information around both sides of the contour. Ten algorithms have been studied that have evolved over a period of time for the segmentation of images. The best performer has been the minimization of region scalable fitting energy for image segmentation approach. It was able to segment the metaspread image with the same accuracy as that of an expert. It is very efficient when it comes to segment the objects that are lying quite close to each other. The only drawback is that it is comparatively slower than other available methods as it results in over segmentation. The selected method is the best candidate for segmenting the metaspread images of human chromosomes.

References

1. Akram F, Kim JH, Lee C, Choi KN (2015) Segmentation of Regions of Interest Using Active Contours with SPF Function. 2015:1–25.
2. Alfredo J, Costa F, De Souza JG (2011) Image Segmentation through Clustering Based on Natural Computing Techniques. Intech 57–82.
3. Ali S, Madabhushi A (2012) An Integrated Region, Boundary, Shape Based Active Contour for Multiple Object Overlap Resolution in Histological Imagery. IEEE Transactions on Medical Imaging 31:1448–1460. doi: 10.1109/TMI.2012.2190089

4. Arachchige AS (2014) Human Metaphase Chromosome Analysis using Image Processing. The University of Western Ontario London, Ontario, Canada
5. Arora T, Dhir R (2014) An Efficient Segmentation Method for Overlapping Chromosome Images. International Journal of Computer Applications 95:29–32.
6. Arora T, Dhir R (2015) A review of metaphase chromosome image selection techniques for automatic karyotype generation. Medical & Biological Engineering & Computing. doi: 10.1007/s11517-015-1419-z
7. Bernard O, Dietenbeck T Creaseg Level-set segmentation platform. Image (Rochester, N.Y.)
8. Bernard O, Friboulet D, Thévenaz P, Unser M (2009) Variational B-spline level-set: a linear filtering approach for fast deformable model evolution. IEEE transactions on image processing : a publication of the IEEE Signal Processing Society 18:1179–1191. doi: 10.1109/TIP.2009.2017343
9. V. CalzadaNavarrete CTH (2014) A Local Adaptive Threshold Approach To Assist Automatic Chromosome Image Segmentation. Latin American Applied Research 44:277–282.
10. Casselles, V., Kimmel R, Sapiro G (1997) Geodesic Active Contours. International Journal of Computer Vision 22:61–79.
11. Castleman HC and ACB and KR (2006) Maximum-likelihood decomposition of overlapping and touching M-FISH chromosomes using geometry, size and color information. "Twenty-Eighth Annual International Conference of the IEEE Engineering in Medicine and Society , New York.
12. Chan T, Vese L (2001) Active Contours Without Edges. IEEE Trans Im Proc 10:266–277.
13. Cremers D, Rousson M, Deriche R (2007) A Review of Statistical Approaches to Level Set Segmentation: Integrating Color, Texture, Motion and Shape. International Journal of Computer Vision 72:195–215. doi: 10.1007/s11263-006-8711-1
14. Devaraj S, Vijaykumar VR, Soundrarajan GR (2013) Leaf Biometrics Based Karyotyping of G-Band Chromosomes. International Journal of Human Genetics 13:131–138.
15. Gagula-Palalic S, Can M (2013) Denver Groups Classification of Human Chromosomes Using CANN Teams Supplemented by a Nearest Neighbor Technique CANN. Southeast Europe Journal of Soft Computing 2:46–51.
16. Grisan E, Poletti E, Ruggeri A (2009) An Improved Segmentation of Chromosomes in Q-Band Prometaphase Images Using a Region Based Level Set. In: World Congress on Medical Physics and Biomedical Engineering,September 7 - 12, 2009, Munich, Germany. pp 748–751
17. Grisan E, Poletti E, Ruggeri A (2009) Automatic Segmentation and Disentangling of Chromosomes in Q-Band Prometaphase Images. IEEE Transactions on Information Technology in Biomedicin 13:575 – 581.
18. Hobart Harris BB (2005) Inside the Cell. Nature. doi: 10.1038/214069a0
19. Jänichen PP and S (2009) Learning of Shape Models from Exemplars of Biological Objects in Images. Transactions on Mass Data Analysis of Images and signals 1:198–209.
20. Joe Hin Tjio and Albert Levan (1925) The Chromosome Number of Man.

Genetics 10:80–85.
21. K MT, Govindan VK (2013) Chromosome Classification using M-FISH Images. In: Proc. of Int. Conf. on Advances in Information Technology and Mobile Communication Chromosome. pp 314–320
22. Karvelis P, Fotiadis DI, Georgiou I, Syrrou M (2006) A Watershed Based Segmentation Method for Multispectral Chromosome Images Classification. In: Proceedings of th 28th IEEE EMBS Annual International COnference , New york,USA, Aug 30-Sept 3,2006. pp 3009–3012
23. Karvelis P, Likas A, Fotiadis DI (2010) Identifying Touching and Overlapping Chromosomes using the Watershed Transform and Gradient Paths. Pattern Recognition Letters 31:2474–2488. doi: 10.1016/j.patrec.2010.08.002
24. Lankton S, Tannenbaum A (2008) Localizing Region-Based Active Contours. IEEE Transactions on Image Processing 17:2029–2039. doi: 10.1109/TIP.2008.2004611
25. Li C, Huang R, Ding Z, Gatenby JC, Metaxas DN, Gore JC (2011) A Level Set Method for Image Segmentation in the Presence of Intensity Inhomogeneities with Application to MRI. IEEE Transactions on Image Processing 20:2007–2016. doi: 10.1109/TIP.2011.2146190
26. Li C, Kao CY, Gore JC, Ding Z (2008) Minimization of region-scalable fitting energy for image segmentation. IEEE Transactions on Image Processing 17:1940–1949. doi: 10.1109/TIP.2008.2002304
27. Liu T, Xu H, Jin W, Liu Z, Zhao Y, Tian W (2014) Medical Image Segmentation Based on a Hybrid Region-Based Active Contour Model. doi: 10.1155/2014/890725
28. Moallem P, Karimizadeh A, Yazdchi M (2013) Using Shape Information and Dark Paths for Automatic Recognition of Touching and Overlapping Chromosomes in G-Band Images. IJ Image, Graphics and Signal Processing 5:22–28. doi: 10.5815/ijigsp.2013.05.03
29. Moradi M, Setarehdan SK, Ghaffari SR (2003) Automatic Locating the Centromere on Human Chromosome Pictures. In: 16th IEEE Symposium Computer-Based Medical Systems, 2003. Proceedings. pp 1–6
30. Mukherjee S, Acton ST (2015) Region based segmentation in presence of intensity inhomogeneity using legendre polynomials. IEEE Signal Processing Letters 22:298–302. doi: 10.1109/LSP.2014.2346538
31. Nobuyuki Otsu (1979) A threshold selection method from gray-level histograms. In: IEEE Trans. Sys., Man., Cyber. pp 62–66
32. Perner, A. Bühring PF and PAGC (Eds. . (2004) Advances in Case-Based Reasoning. In: Proceedings of the ECCBR 2004, Madrid/Spain, Springer Verlag. pp 375–388
33. Piper J, Granum E (1989) On Fully Automatic Feature Measurement for Banded Chromosome Classification. Cytometry 10:242–255.
34. Poletti E, Zappelli F, Ruggeri A, Grisan E (2012) A review of thresholding strategies applied to human chromosome segmentation. Computer Methods and Programs in Biomedicine 108:679–688. doi: 10.1016/j.cmpb.2011.12.003
35. Schwartzkopf WC, Bovik AC, Evans BL (2005) Maximum-Likelihood Techniques for Joint Segmentation-Classification of Multispectral

Chromosome Images. IEEE Transactions on Medical Imaging 24:1593–1610. doi: 10.1109/TMI.2005.859207
36. Shi Y, Karl WC, Member S (2008) A Real-Time Algorithm for the Approximation of Level-Set-Based Curve Evolution. 17:645–656. doi: 10.1109/TIP.2008.920737
37. Somasundaram D, Nirmala M (2010) Automatic Segmentation And Karyotyping Of Chromosomes Using Bio-Metrics. In: International Conference on Emerging Trends in Robotics and Communication Technologies (INTERACT). IEEE, pp 42–45
38. Somasundaram D, Kumar VRV (2014) Separation of Overlapped Chromosomes and Pairing of Similar Chromosomes for Karyotyping Analysis. Measurement 48:274–281. doi: 10.1016/j.measurement.2013.11.024
39. Somasundaram D, Palaniswami S, Vijayabhasker R, Venkatesakumar V (2014) G-Band Chromosome Segmentation, Overlapped Chromosome Separation and Visible Band Calculation. Int J Hum Genet 14:73–81.
40. Sreejini K S LA and VKG (2012) M-FISH Karyotyping- A New Approach Based On Watershed Transform. International Journal of Computer Science, Engineering and Information Technology (IJCSEIT) 2:105–117.
41. Stanley RJ, Keller J, Caldwell CW, Gader P (1996) Centromere Attribute Integration Based Chromosome Polarity Assignment. In: conference of the American Medical Informatics Association. pp 284–8
42. Suri JS, Liu K, Singh S, Laxminarayan SN, Zeng X, Reden L (2002) Shape recovery algorithms using level sets in 2-D/3-D medical imagery: A state-of-the-art review. IEEE Transactions on Information Technology in Biomedicine 6:8–28. doi: 10.1109/4233.992158
43. Uttamatanin R, Yuvapoositanon P, Intarapanich A, Kaewkamnerd S, Phuksaritanon R, Assawamakin A, Tongsima S (2013) MetaSel: A Metaphase Selection Tool using a Gaussian-Based Classification Technique. BMC Bioinformatics 14:S13. doi: 10.1186/1471-2105-14-S16-S13
44. Yang, J. D., Chen, Y. S., Hsu WH (1994) Adaptive thresholding algorithm and its hardware implementation. Pattern Recognition Letters 15.:141–150.
45. Zhang K, Zhang L, Song H, Zhou W (2010) Active contours with selective local or global segmentation: A new formulation and level set method. Image and Vision Computing 28:668–676. doi: 10.1016/j.imavis.2009.10.009
46. M-FISH database established by Advanced Digital Imaging Research (http://www.adires.com/05/Project/ MFISH_DB/MFISH_DB.shtml.

Extraction Kinetics of Ultrasound-assisted Extraction of Oleanolic Acid from *Hedyotis corymbosa*

Ming-Chi Wei[1], Pei-Hui Lin[2], Show-Jen Hong[3], and Yu-Chiao Yang[3*]

[1] Department of Applied Geoinformatics, Chia Nan University of Pharmacy and Science,Tainan 71710, Taiwan
[2] Department of Surgery, Davis Heart and Lung Research Institute,The Ohio State University Wexner Medical Center, Columbus, Ohio 43210, USA
[3] Department and Graduate Institute of Pharmacology, Kaohsiung Medical University, Kaohsiung 80708, Taiwan

[1]Ycyang230@gmail.com

Abstract. This study investigated a hyphenated procedure of heat-reflux and ultrasound-assisted extraction (HUAE) to improve the extraction efficiency of classic solvent extraction techniques, such as heat-reflux extraction, to extract oleanolic acid (OA) from *Hedyotis corymbosa*. The superior yield was achieved at a a mean particle size of 0.355 mm, a duty cycle of 75%, a liquid/solid ratio of 12 mL/g, an ethanol concentration of 80% and with extraction temperatures of 56 °C. Compared with heat-reflux extraction method, the HUAE technique reduced the extraction time, extraction temperature and solvent consumption and afforded a higher extraction yield of the target compound from the plant matrix. A second-order kinetic model described the extraction process under different extraction temperatures and showed good agreement with the experimental results.

Keywords: *Hedyotis corymbosa*, oleanolic acid, heat-reflux extraction, ultrasound-assisted extraction.

1 Introduction

Hedyotis corymbosa (L.) Lam. (synonym *Oldenlandia corymbosa* (L.) Lam.) belongs to an evergreen Rubiaceae genus, and the medicinal herb is derived from the dried whole plants. The herb is widely used in China as a healthy tea [1] and as a traditional Chinese herbal medicine for the treatment of various diseases, such as skin disease, malaria, intestinal abscess, hepatitis and cancer [2]. Oleanolic acid (OA) is a natural pentacyclic triterpenoid carboxyl acid and is one of the most famous bioactive plant compounds (Fig. 1). Many plants contain OA and have been reported to exhibit a wide range of biological functions, including anti-cancer, chemopreventive, hepatoprotective, antiviral, antibacterial, anti-diabetic, antioxidant, anti-inflammatory and gastroprotective properties [3,4]. Furthermore, OA can inhibit proliferation and induce apoptosis in many types of cancer cells, including hepatocellular carcinoma, prostate carcinoma, colorectal cancer, acute myelogenous leukaemia, skin tumourigenesis, cervical carcinoma and lung carcinoma [5].

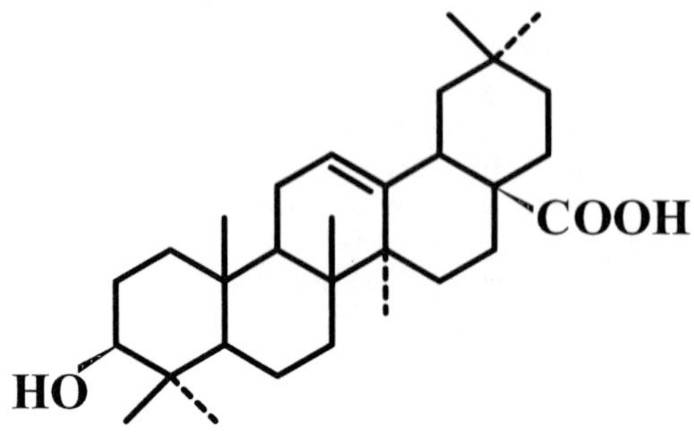

Fig. 1. Chemical structure of oleanolic acid (OA).

Several extraction techniques have been used to extract OA from plant matrix, including Soxhlet extraction, maceration [6], heat-reflux extraction (HRE) [7], shaking [8] (Yang and Wei, 2015b), ultrasound-assisted extraction (UAE) [9], microwave extraction [10] and supercritical carbon dioxide extraction (SC-CO_2) [11]. Ultrasound-assisted extraction (UAE) is a potential alternative technology to the conventional techniques due to its reduced cost and instrumental requirements. Some characteristics of UAE are high extraction efficiency, good reproducibility, simplified manipulation, low consumption of organic solvents and time and lower energy input. Ultrasonic enhancement during extraction is attributed to the disruption of cell walls, particle size reduction, improved penetration, enhanced swelling and hydration processes and enhanced mass transfer of the cell content via cavitation bubble collapses [9]. Furthermore, the combination of energy sources by using conventional extractions under ultrasound irradiation can prevent the destruction of target components and improve the extraction of natural products. Thus, ultrasound extraction can be combined with the other extraction procedures, such as heat-reflux extraction (HRE) [7], SC-CO_2 extraction [11] and microwave-assisted extraction [12], to improve efficiency and yield. Previous studies [7] have shown that the simultaneous application of UAE and HRE can significantly improve the extraction efficiency, require less solvent, shorten the extraction time and extract thermally labile compounds under mild conditions compared with either operation. However, no references have been found in the literature reporting the simultaneous application of UAE and HRE for the extraction of OA from *H. corymbosa*.

The extraction yield and efficiency of the target compounds from raw materials are strongly influenced by the operating conditions during extraction. For solid-liquid extraction processes, the kinetics of solute extraction from raw sources involves releasing the solute from porous matrices into the solvent phase by mass transfer. Therefore, understanding mass transfer at the solid-liquid interface plays a significant role in scaling up the extraction process. However, no relevant kinetic model of ultrasound-assisted extraction of OA from *H. corymbosa* has been reported. Therefore, the objective of this work is to outline the feasibility of a hyphenated technique of HRE and UAE (HUAE) in the fast preparation of high-yield extracts rich in OA from *H. corymbosa*. Several parameters that could potentially affect the extraction efficiency were evaluated. Finally, the experimental kinetic data curves were validated using a second-order model for predicting the extraction processes and revealing the extraction mechanism.

2 Materials and methods

2.1 Raw material

Three sets of authentic plants of dried whole *H. corymbosa* (Sample HC1 to HC3) were procured from different local Chinese medicinal shops (Kaohsiung, Taiwan). All materials were identified and authenticated by professor Ming-Chi Wei with voucher specimens (Sample HC1 to HC3) deposited in the Department and Graduate Institute of Pharmacology, Kaohsiung Medical University (Kaohsiung, Taiwan). All samples were triturated in a knife mill, and were separated according to their sizes (standard testing sieve, series Tyler) with different mean particle sizes of 0.93, 0.73, 0.55, 0.36 and 0.11 mm. Subsequently, the moisture content (% dry weight basis) was determined using oven method for which the methodology has been discussed in detail elsewhere [7]. Briefly, the moisture content (% dry weight basis) was determined by drying at 105 °C to a constant mass and was 11.68, 13.21 and 10.93% for samples HC1 to HC3, respectively. All the yields were calculated based on a moisture-free basis and represent the mean values of at least six experiments. In the study, all the experiments for *H. corymbosa* were performed on sample HC2.

2.2 Chemicals and reagents

Oleanolic acid (95%) is purchased as HPLC reference standards from Sigma Chemical Co. (St. Louis, MO, USA) and used without further purification. Methanol (99.9%), ethanol (99.9%), acetone (99.7%), acetonitrile (99.9%), ethyl acetate (99.9%), *n*-hexane (95%) and phosphoric acid (85%) were bought from Merck Co. (Darmstadt, Germany). Cyclohexane and sulfuric acid were supplied by Tedia (Fairfield, OH, USA) and Acros Organics (Morris Plains, NJ, USA), respectively. The deionized water was purified using a Milli-Q reverse osmosis unit (Millipore, Bedford, MA, USA).

2.3 Hyphenated heat-reflux and ultrasound-assisted extraction (HUAE) procedure

The HUAE procedure used in this work has been discussed in detail elsewhere [8] and is briefly described here. The experimental set-up consisted of two extraction stages: HRE (stirring at 300 rpm) and UAE (without stirring). For each experiment, a 5 min HRE was followed by a 2.5–30 min UAE. During the HUAE, 5 g of dried, powdered plant was placed in a 250 mL conical flask with the appropriate extraction solvent volume; this flask was immersed in a temperature-controlled ultrasonic bath (40 kHz working frequency and 185 W power, Branson B-33510E-DTH, USA) with stirring (immersion-stirring device, 300 rpm; Thermo Scientific Variomag Compact and Maxi, USA). The desired extraction temperature was adjusted, and a 5 min HRE was then used to obtain good solvent-to-plant material contact. The water in the ultrasonic bath was circulated and regulated at a constant temperature (Haake F3-K, Haake, Karlsruhe, Germany) to prevent any temperature increase from the ultrasonic waves. To avoid evaporating the extracting solvent, a condenser was connected to the conical flask. After the HRE, the immersion-stirring device was removed from the ultrasonic cleaning bath; UAE was then performed using an intermittent pulse mode. The UAE cycle time consisted of a pulse duration period and a pulse interval period. The ultrasound exposure duty cycle (intermittent sonication, expressed as a percentage) was the proportion of the cycle time spent as a pulse.

The HUAE experiments were performed using various solvent to raw material ratios (6–20 mL/g), duty cycles (0–100%), ethanol concentrations (0–99.5%, v/v), particle sizes (0.925, 0.725, 0.550, 0.355 and 0.106 mm), number of extraction cycles (1–3) and extraction temperatures (30–65 °C). The extraction times ranged from 2.5 to 30 min. All extractions were performed six times for each sample. After the contact time elapsed, the flask contents were immediately filtered before concentration using a rotary evaporator (Eyela Rotary Evaporator N-1000, Tokyo Rikakikai Co., Ltd, Japan).

2.4 Heat-reflux extraction (HRE)

OA was subjected to HRE for different durations with a stirring speed of 300 rpm; this process was treated as the control. The HRE of OA from *H. corymbosa* was performed in a 250 mL conical flask using the same methods as previous described [11]. Six replicate experiments were performed with various extraction temperatures (30–80 °C), durations (30–120 min), solvent to material ratios (6–20 mL/g), ethanol concentrations (0–99.5%, v/v), mean particle sizes (0.925, 0.725, 0.550, 0.355 and 0.106 mm) and extraction cycles (1–4).

2.5 High-performance liquid chromatography (HPLC)

The extracts obtained from best HUAE condition and conventional extraction techniques were analyzed using a Jasco HPLC system which was composed of a column thermostat, a solvent degasser, a Jasco PU-986 intelligent HPLC pump, a Jasco MD-910 intelligent ultraviolet/visible (UV/vis) multi-wavelength detector, and a Jasco AS-2057 plus intelligent auto sampler (Jasco, Tokyo, Japan). Jasco Borwin 1.21 chromatography software (Jasco, Tokyo, Japan) was used for the manipulation of data processing.

All analyses were carried out using initial conditions chosen according to the literatures [9] and were modified to our purpose as follows. Chromatographic separation was carried out in a reverse-phase LiChrospher® C-18 column (250 mm × 4 mm i.d., 5-µm particle size) connected to a LiChroCart guard column containing the same packing material (both Merck, Darmstadt, Germany) at 40 °C, using a mobile phase consisting of acetonitrile (solvent A) and 0.1% phosphoric acid in water (v/v) (solvent B) under the following gradient profile: 0–26 min, 22–24% (solvent A), flow rate of 1.0–

1.5 mL/min; 26–42 min, 24% (solvent A), flow rate of 1.5–1.0 mL/min; 42–60 min, 24–92% (solvent A), flow rate of 1.0 mL/min. The injection volume for all samples was 20-µL, and UV detection was carried out at 210 nm for OA. Identification of OA was performed by comparing the retention times and UV–visible spectra of peaks in extracts to those of OA standards. Quantification of OA was achieved by the absorbance recorded in the chromatograms relative to external standards. The extraction yield of OA was calculated from peak area according to calibration curves and was expressed as milligrams per gram of dry plant (mg/g DW).

2.6 Statistical Analysis

All yields and composition analyses were calculated based on a moisture-free basis. The mean and standard deviation (SD) of the mean were calculated from six experiments. The results are expressed as the mean ± SD. Analysis of variance (ANOVA) was carried out using Tukey's method with a significance level of $P < 0.05$ using 2010 Microsoft Office Excel (Microsoft Co., USA) and Origin software version 6.1 (Origin Lab Co., Northampton, MA, USA). Furthermore, the concordance between experimental data and calculated value was established by the average absolute relative deviation (*AARD*) and the coefficient of determination (R_2), which were established using the following equations:

$$AARD(\%) = \frac{100}{n}\sum_{i=1}^{n}\left|\frac{y_{p,i} - y_{e,i}}{y_{e,i}}\right| \qquad (1)$$

$$R^2 = \frac{\sum_{i=1}^{n}(y_{p,i} - y_{e,i})^2}{(y_{p,i} - y_m)^2} \qquad (2)$$

where $y_{p,i}$ and $y_{e,i}$ are the predicted and experimental values of the extraction yield, y_m is the mean value of the extraction yield, and n is the number of experimental runs.

3 Results and discussions

3.1 Result of HPLC analysis of samples

The HPLC chromatograms of the OA standard and the aqueous ethanol extracts of HRE and HUAE are shown in Fig. 2. Fig. 2 A shows the standard substance with a retention time of 57.17 min for OA. Figs. 2 B and C show chromatograms of dried *H. corymbosa* HRE and HUAE extracts, respectively, and the other compounds in the samples did not interfere with the analysis of OA at the elution time of the analyte (57.17 min). Peaks in the chromatograms obtained from *H. corymbosa* samples were identified by comparing the retention time with an authentic standard (OA). Therefore, the HPLC system successfully separated and identified OA in the *H. corymbosa* extracts (Figs. 2 B and C) and was used to quantify the content of OA. Therefore, OA was quantified in *H.*

corymbosa extracts using a regression equation of the calibration curve of pure OA expressed as milligrams per gram of dry weight (mg/g of DW).

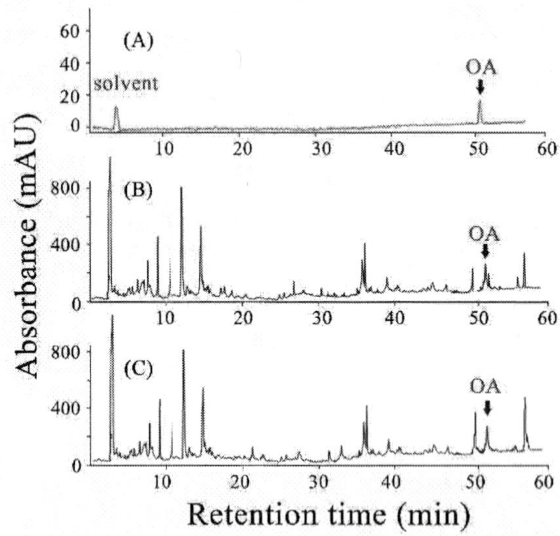

Fig. 2. Typical HPLC chromatograms obtained from a standard solution of OA (A) and *H. corymbosa* extracts obtained from HRE (B) and HUAE (C).

3.2 Effect of extraction temperature

The effects of extraction temperature on the extraction yield of OA were determined at a mean particle size of 0.355 mm, a duty cycle of 75%, a liquid/solid ratio of 12 mL/g, an ethanol concentration of 80% and various extraction times (7.5 to 35 min) with extraction temperatures of 31, 37, 44, 50, 56 and 65 °C, as shown in Fig. 3. The extraction yield of OA increased with increasing extraction temperature from 31 to 56 °C and approached a peak value at 56 °C due to the increased diffusion coefficient and increased solubility of OA. However, when the extraction temperature increased from 56 °C to 65 °C, the extraction yield started to decrease due to the increased volatility of UA at higher temperatures. The main effect of ultrasonication was cavitational bubble collapse. The effects of temperature appeared to be due to acoustic cavitation because higher efficiency was observed at 56 °C compared with 65 °C. At a lower temperature of 56 °C, the number of cavitation bubbles increases, and their collapse is powerful enough to effectively extract OA from *H. corymbosa.* Although the number of bubbles is higher, their collapse is less efficient at 65 °C, which causes the cavitational effects to exhibit a stronger intensity at 56 °C compared with 65 °C. Additionally, increasing extraction temperature might result in increased solvent volatilization, improved energy cost and enhanced impurity extraction; therefore, 56 °C was selected as the optimum temperature to obtain the highest extraction yield of OA.

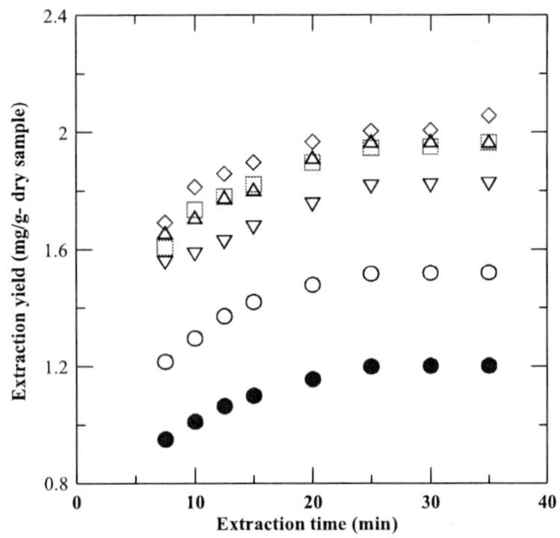

Fig. 3. Influence of temperature on the extraction yield of OA from H. corymbosa using HUAE process. Extraction conditions: duty cycle: 75%; ethanol concentration: 80% (v/v); mean particle size: 0.355 mm; ratio of solvent to raw material: 12:1 (mL/g). (●: 31 °C, ○ : 37 °C, ▽: 44 °C, △ :50 °C, ◇ :56 °C, □ :65 °C).[1]UA yield (mg/g) = weight of the extracted UA/ weight of feeding material.

3.3 Kinetics of hyphenated heat-reflux and ultrasound-assisted extraction

The typical kinetic models of solid–liquid extractions, such as Soxhlet extraction and heat-reflux extraction, were appropriately fitted using a second-order rate law [7] to experimentally evaluate the extraction rate constant and concentration at saturation. Because the HUAE process exhibited a similar shape compared with that of the solid–liquid extraction process and the enhancements by the cavitation effects of ultrasonic waves, the curves could be described by the same model [9]. According to a second-order rate law, the rate of dissolution of the OA from the plant cells into solution can be written as:

$$\frac{dC}{dt} = k(C_s - C)^2 \qquad (3)$$

Where k is the second-order extraction rate constant (L/g min), C is the concentration of target compounds in the liquid extract at a given extraction time t (mg OA /g dry plant) and Cs is the extraction capacity (concentration of OA at saturation in mg OA /g dry plant).

The integrated rate law for a second-order extraction under the initial and boundary conditions $t = 0$ to t and $C = 0$ to C, C was obtained from the following equations:

However, the saturated extraction capacity (*Cs*) and the initial extraction rate (*h*) all increased with temperature, which agreed with previous reports [7].

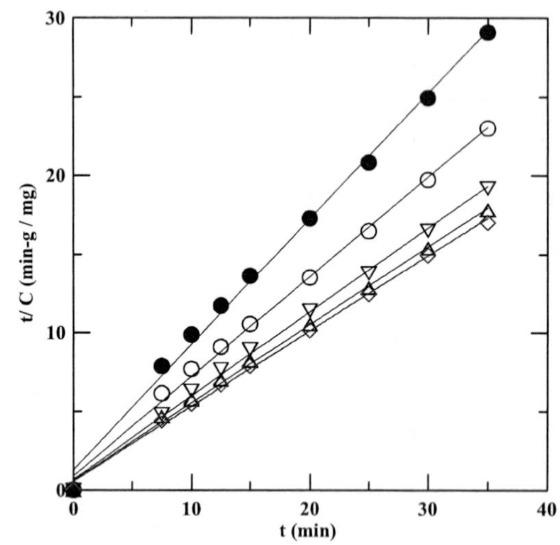

Fig. 4. Second-order extraction kinetics of OA from H. corymbosaat at various temperatures using the HUAE process. (●: 31 °C, ○ : 37 °C, ▽: 44 °C, △ :50 °C, ◇ :56 °C,).

$$\int_0^C \frac{dC}{(C_s - C)^2} = \int_0^t kdt \tag{4}$$

$$C = \frac{C_s^2 kt}{1 + C_s kt} \tag{5}$$

By rearranging Eq. (5), its linear form would be as follows:

$$\frac{t}{C} = \frac{1}{kC_s^2} + \frac{t}{C_s} \tag{6}$$

When t approaches 0, the initial extraction rate, h, can be defined as follows:

$$h = kC_s^2 \tag{7}$$

The values for *h*, *Cs* and *k* were determined experimentally using the slope and intercept by plotting *t/C* versus *t*.

Based on Eq. (6), the t/C values were plotted over time, and the results were fitted as a straight line (Fig. 4) by a linear least squares regression method to estimate Cs and k. The values Cs and k were used to evaluate h values. However, because a slight decrease in the extraction yield was observed when increasing the temperature from 55 to 65 °C (Fig. 3), the experimental kinetic data curve at 65 °C was not included in Fig. 4. The parameters for the second-order rate law, the coefficient of determination (R_2) and the overall average absolute relative deviation (*AARD*) are listed in Table 1. The straight-line curve is shown in Fig. 4, and the high correlation coefficients at all of the temperatures (R_2 >0.996) and the low overall *AARD* (%) imply good agreement of the second order extraction model with the experimental results. Table 1 also showed that the extraction rate constant (k) decreased with increasing extraction temperature.

Table 1. Parameters of kinetic model for the extraction of OA from H. corymbosaat using HUAE technique..

Compounds	T (°C)[a]	h (mg/g-min)[b]	k (g/mg-min)[c]	Cs (mg/g)[d]	R_2 [e]
OA	31	0.775	0.493	1.254	0.995
	37	1.110	0.445	1.579	0.996
	44	1.482	0.421	1.877	0.997
	50	1.633	0.406	2.020	0.997
	56	1.769	0.400	2.093	0.998

[a] Extraction temperature.
[b] Initial extraction rate.
[c] Rate constant.
[d] Equilibrium concentration.
[e] Regression coefficient.

4 Conclusions

To effectively extract value-added oleanolic acid from *H. corymbosa*, the effects of HUAE processing parameters such as time and temperature on extraction performance and global kinetics were systematically studied. The extraction kinetics were investigated by varying time (7.5 to 35 min) and temperature (31 to 65 °C). The results showed that HUAE offers a promising alternative for the efficient extraction of OA from plant matrix. Furthermore, a second-order kinetics model was in good agreement with the experimental results.

References

1. Zhao, J., Deng, J.W., Chen, Y.W., Li, S.P.: Advanced Phytochemical Analysis of Herbal Tea in China. J. Chromatogr. A 1313 (2013) 2–23.
2. Moniruzzaman, M., Ferdous, A., Irin, S.: Evaluation of Antinociceptive Effect of Ethanol Extract of *Hedyotis Corymbosa* Linn. Whole Plant in Mice. J. Ethnopharmacol. 161 (2015) 82–85.

3. Alvarado, H.L., Abrego, G., Souto, E.B., Garduno-Ramirez, M.L., Clares, B., Garcia, M.L., Calpena, A.C.: Nanoemulsions for Dermal Controlled Release of Oleanolic and Ursolic Acids: In Vitro, Ex Vivo and In Vivo Characterization. Colloids Surf. B Biointerfaces 130 (2015) 40–47.
4. Liese, J., Abhari, B.A., Fulda, S.: Smac Mimetic and Oleanolic Acid Synergize Toinduce Cell Death in Human Hepatocellular Carcinoma Cells. Cancer Letters 365 (2015) 47–56.
5. Bednarczyk-Cwynar, B., Ruszkowski, P., Bobkiewicz-Kozlowska, T., Zaprutko, L.: Oleanolic Acid A-lactams Inhibit the Growth of HeLa, KB, MCF-7 and Hep-G_2 Cancer Cell Lines at Micromolar Concentrations Anticancer Agents Med Chem. 16 (2016) 579–592.
6. Magdalena, W.K., Ireneusz, S., Ryszard, K., Renata, N.: Effect of Different Extraction Techniques on Quantification of Oleanolic and Ursolic Acid in *Lamii Albi* Flos. Industrial Crops and Products 44 (2013) 373– 377.
7. Yang, Y.C., Wei, M.C.: Kinetic and Characterization Studies for Three Bioactive Compounds Extracted from *Rabdosia rubescens* Using Ultrasound. Food Bioprod. Process. 94 (2015) 101–113.
8. Wei, M.C., Yang, Y.C.: Kinetic Studies for Ultrasound-Assisted Supercritical Carbon Dioxide Extraction of Triterpenic Acids from Healthy Tea Ingredient *Hedyotis diffusa* and *Hedyotis corymbosa*. Sep. Purif. Technol. 142 (2015) 316–325.
9. Yang, Y.C., Wei, M.C.: A Combined Procedure of Ultrasound-assisted and Supercritical Carbon Dioxide for Extraction and Quantitation Oleanolic and Ursolic Acids from *Hedyotis corymbosa*. Ind. Crops Prod. 79 (2016) 7–17.
10. Verma, S.C., Jain, C.L., Kumari, A., Padhi, M.M., Devalla, R.B.: Microwave-assisted Extraction and Rapid Isolation of Ursolic Acid from the Leaves of Eucalyptus × Hybrida maiden and Its Quantification Using HPLC-Diode Array Technique. J. Sep. Sci. 36 (2013) 1255–1262.
11. Wei, M.C., Xiao, J., Yang, Y.C.: Extraction of A-Humulene-Enriched Oil from Clove Using Ultrasound-assisted Supercritical Carbon Dioxide Extraction and Studies of Its Fictitious Solubility. Food Chemistry, 210 (2016) 172–181.
12. Leonelli, C., Mason, T.J.: Microwave and ultrasonic processing: now a realistic option for industry. Chemical Engineering and Processing 49 (2010) 885–900.

Ultrasound-assisted Supercritical Carbon Dioxide Extraction of Ursolic Acid from *Rabdosia rubescens*

Yu-Chiao Yang[1], Pei-Hui Lin[2], Jin-Ming Chen[3], Yu-Tuan Chou [3], and Ming-Chi Wei[3*]

[1] Department and Graduate Institute of Pharmacology,
Kaohsiung Medical University, Kaohsiung 80708, Taiwan
[2] Department of Surgery, Davis Heart and Lung Research Institute,
The Ohio State University Wexner Medical Center, Columbus, Ohio 43210, USA
[3] Department of Applied Geoinformatics,
Chia Nan University of Pharmacy and Science,Tainan 71710, Taiwan
mcwei2430@gmail.com

Abstract. Ursolic acid (UA) exhibits an antineoplastic activity on several kinds of cancer. In this study, the temperature of 32–64 °C and pressure of 11.0–32.0 MPa were used to extract UA from *Rabdosia rubescens* using ultrasound-assisted supercritical carbon dioxide (USC–CO_2) with ethanol solution as a cosolvent. The optimized condition for maximal extraction yield of UA using USC–CO_2 process was at a pressure of 28.5 MPa and temperature of 55 °C with 11.6% of ethanol (80%, v/v) as cosolvent, leading to the extraction yield of 2.386 mg/g dry plant weight. Comparing with other conventional methods, USC–CO_2 extraction revealed superior extraction yield and efficiency. In conclusion, USC–CO_2 procedure can be considered as a promising green alternative to extract UA from *R. rubescens* for food, cosmetic and pharmaceutical industries with benefits of no chemical waste, rich in active compounds and short extraction time.

keywords: *Rabdosia rubescens*, ursolic acid, ultrasound-assisted supercritical CO_2 extraction, cosolvent.

1 Introduction

The entire dried *Rabdosia rubescens* (Hemsl.) Hara plant (known as Dong-ling-cao in China) is the most popular healthy teas throughout Asia for health care and diseases prevention worldwide [1]. Traditionally, *R. rubescens* is used for the treatment of stomachache, pharyngitis, sore throat, cough, wrestling injuries and tumors in Chinese folk medicine. Recently, this herb has gained attention to its anticancer activates [2]. Ursolic acid (UA, 3β-hydroxy-urs-12-en-28-oic acid) (Fig. 1) is one important constituent of *R. rubescens* and is a common constituent of medical herbs and edible plants, which widely applied in medicine with many

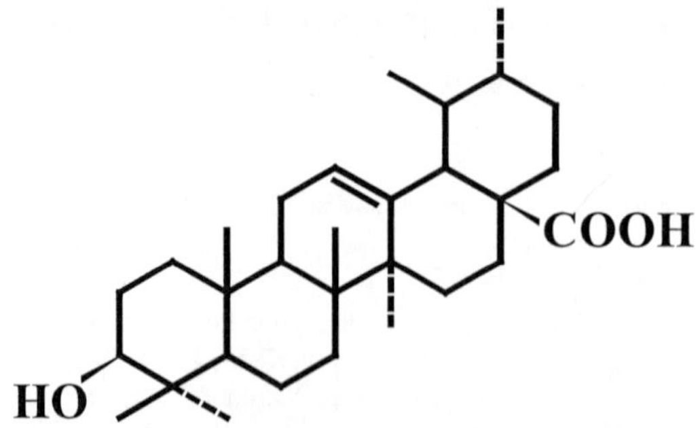

Fig. 1. Chemical structure of ursolic acid (UA).

beneficial functions to human health. In recent years, the literature furnishes numerous data on UA induce apoptosis in a wide variety of cancer cells, including hepatocellular carcinoma, prostate carcinoma, colorectal cancer, acute myelogenous leukemia, skin tumorigenesis, cervical carcinoma and lung carcinoma [3]. Therefore, UA is an important anticancer constituent and contribute to the pharmacological efficacy of the herb.

Early works have reported the extraction of UA in other natural materials using supercritical carbon dioxide (SC–CO_2) [4,5] and traditional solvent extraction method [6]. However, the traditional solvent extraction methods are usually characterized by the consumption of large volumes of solvent and energy, low extraction yields, lengthy extraction procedures and the potentially deleterious degradation of labile compounds. SC–CO_2 extraction has capabilities to overcome the limitations [7]. When CO_2 is pushed beyond its critical point (31.1 °C and 7.38 MPa), CO_2 has been recognized as an ideal non-polar extraction solvent, because SC–CO_2 exploits the unique properties of materials, including high dissolving power, high diffusivity, and low viscosity. More importantly, CO_2 is nontoxic, and can be readily recovered after extraction. Because CO_2 is considered to be very safe for humans, SC–CO_2 extraction process is an acceptable environmentally friendly alternative. Furthermore, the extracts obtained using SC–CO_2 extraction are of superior quality as compared to those obtained using the conventional organic solvent extraction methods. Therefore, SC–CO_2 extraction has become an important separation technique in the field of food, pharmaceutical and cosmetic industries.

With the development of SC–CO_2 extraction, it is found out that polar analytes are poorly soluble in SC–CO_2 due to its lack of polarity. Thus, pure SC–CO_2 frequently fails to efficiently extract many organics from a sample matrix. However, the solubility of the polar compounds in SC–CO_2 can be enhanced by adding cosolvents (or modifiers). The introduction of organic cosolvents is an effective method to broaden the application of SC–CO_2 extraction. For example, Wei and Yang [8] report ultrasound-assisted SC–CO_2 extraction of oleanolic acid (OA) and UA from *Hedyotis diffusa* under a temperature range of 30–63 °C and a pressure range of 10.4–30 MPa. They used aqueous ethanol as a cosolvent to improve the solubilities of OA and UA in SC–CO_2 and to make these processes feasible. Similarly, Domingues et al. [4] used 5% aqueous ethanol as a cosolvent to extract triterpenic acids from *Eucalyptus globulus* bark using SC–CO_2 extraction, and Yang et al. [9] used 11.9% ethanol solution (80%, v/v) as a cosolvent to extract OA and UA from *Scutellaria barbata* D. Don using ultrasound-assisted SC–CO_2 extraction.

On the other hand, the simultaneous application of SC–CO_2 and ultrasound has been shown to improve the traditional SC–CO_2 extraction processes [10]. In the ultrasound-assisted SC–CO_2 (USC–CO_2) extraction processes, the ultrasound-induced micro-stirring and solvent cavitation had some physical consequences, including cracked or damaged plant cell walls, increased solvent diffusion, interfacial turbulence and a reduction of the external resistance to mass transfer, thus improving the mass transfer of the solvents into the raw materials and the soluble constituents into the solvents. Recently, USC–CO_2 extraction was shown to significantly decrease extraction time and increase extraction yields in natural materials while utilizing less severe operating parameters, such as temperature, pressure, CO_2 flow rate. However, there are almost no previous reports regarding the USC–CO_2 extraction process of UA from *R. rubescens*. Therefore, the main objective of this study was to apply an USC–CO_2 extraction process for the preparation of UA-enriched extracts from *R. rubescens* and to compare these extracts with those obtained from a conventional solvent extraction technique. The novel method consists of static extraction (with ultrasound-assisted), followed by SC–CO_2 dynamic extraction (with ultrasound-assisted), which is the first time that this novel method has been investigated. The ultrasound-assisted SC–CO_2 extraction of UA from *R. rubescens* was performed at several temperatures (32, 39, 47, 55, and 64 °C) and pressures (11.0, 18.5, 23.5, 28.5 and 32.0 MPa) with ethanol solution as a cosolvent.

2 Materials and methods

2.1 Plant materials

Five samples of *R. rubescens* (Sample RR1 to RR5) were purchased from different local Chinese medicinal shops (Taiwan). All the samples were sorted and identified by the Department and Graduate Institute of Pharmacology, Kaohsiung Medical University (Kaohsiung, Taiwan) and the quality of the samples complied with the Chinese Pharmacopoeia.

2.2 Solvents and reagents

CO_2 was purchased in the liquid form from Yun-Shan Gas Co. Ltd. (Tainan, Taiwan). UA was procured as HPLC reference standards from the Sigma Chemical Co. (St.

Louis, MO, USA). Methanol, ethanol, acetone, acetonitrile, ethyl acetate, *n*-hexane and 85% phosphoric acid were bought from Merck Co. (Darmstadt, Germany).

2.3 Ultrasound-assisted supercritical carbon dioxide (USC–CO_2) extraction

All experiments were conducted using ultrasound-assisted extraction (static extraction) with a 15 min duration, followed by SC–CO_2 extraction (with ultrasound-assisted). The static stage lasted 15 min for all experiments, while the dynamic stage varied from 10 to 150 min (with ultrasound-assisted). More details about the equipment and its operation have recently been described elsewhere [11]. For each experiment, a mixture of dried and ground herb (20 g, 0.36 mm) and glass beads (1 mm diameter) was placed in a 152 mL stainless steel extraction vessel (SS304, i.d. of 2.2 cm and length of 40.0 cm)..

2.4 Other extraction methods

Various extraction methods [12], including heat-reflux extraction (HRE), and conventional SC–CO_2 extraction, were also investigated and compared to the USC–CO_2 extraction process..

2.5 HPLC analysis

HPLC analysis of UA was carried out on a Jasco HPLC system with a LiChrospher® C-18 analytical column (250 mm × 4 mm i.d., 5-μm particle size). The mobile phase was composed of acetonitrile (A) and 0.1% phosphoric acid (B). More details about the equipment and its separation conditions can be found in a previous work [13].

2.6 Statistical Analysis

All yields were calculated based on a moisture-free basis. The mean and standard error of the mean were calculated from six experiments. The results are expressed as the mean ± SD. Analysis of variance (ANOVA) was carried out using Tukey's method with a significance level of $P < 0.05$ using 2010 Microsoft Office Excel (Microsoft Co., USA) and Origin version 6.1 software (Origin Lab Co., Northampton, MA, USA).

3 Results and discussions

3.1 USC–CO_2 extraction of UA from *Rabdosia rubescens*

Because SC–CO_2 is a non-polar substance, SC–CO_2 often fails to extract polar organic compounds from raw matrices. The addition of polar cosolvent was necessary to overcome the solubility limitations of pure SC–CO_2, which allowed the extraction of more polar substances from target samples. Yang and Wei [14] found that the addition of 11.6% (v/v) aqueous ethanol (ethanol/water = 80/20, v/v) was optimum to obtain the highest OA and UA yield from *H. diffusa*. The result supported previous findings that ethanol/water was the best cosolvent for the extraction of triterpene acids from *S. barbata* D. Don in SC–CO_2 [9]. One important effect is that cosolvent improves the swelling of matrix, which may affect the cell structure, improves the diffusion of target compounds from matrix and enhances the dissolution into the SC–CO_2. At the initial stages of extraction, the interaction of cosolvent and matrix is anticipated to be greater. Then it should be stabilized so that the concentration of cosolvent in the system is constant for the rest period. The complex interactions among the target matrix, target compounds and SC–CO_2 lead to the higher amounts of OA and UA being extracted. Therefore, the subsequent experiments were carried out with 11.6% (v/v) aqueous ethanol (ethanol/water = 80/20, v/v) as cosolvent for extraction of UA from *R. rubescens*.

For SC–CO_2 extraction, the extraction pressure and temperature are two main parameters that influence the extraction efficiency and selectivity. In this study, the effects of extraction pressure and temperature on the extraction yield of UA were performed for 75 min dynamic stage (with ultrasound-assisted) at a mean particle size of 0.36 mm, a static extraction time of 15 min (with ultrasound-assisted), a SC–CO_2 flow rate of 2.1 mL/min (STP) and a cosolvent (80% aqueous ethanol) percentage of 11.6% (v/v). The ultrasound for static and dynamic stages was fixed at an ultrasonic frequency of 40 kHz, a power of 185 W and an ultrasound cycle of 75% (intermittent sonication). Fig. 2 shows the effect of various temperatures (32, 39, 47, 55 and 64 °C) and pressures (11.0, 18.5, 23.5, 28.5 and 32.0 MPa) on the extraction yield of UA form *R. rubescens* with ethanol:water (80:20, v/v) as cosolvent. The extraction yield of UA enhanced significantly with increase of pressure (11.0–28.5 MPa) at the operating temperatures. These results were predictable because raising the extraction pressure, at constant temperature, leads to a higher density of SC–CO_2, which increases the solubility of UA. A similar outcome was reported by Patinha et al. [5], who found the triterpenic acids can significantly extracted at 20 MPa and 60 °C using SC–CO_2. The results were also consistent with those reported by de Melo et al. [15] and Domingues et al. [4] using SC–CO_2 to extract triterpenic acids from *Eucalyptus globulus* bark. However, as the pressure increases from 28.5 to 32.0 MPa at various temperatures, the increase of pressure did not have any significant effect on the extraction yield of UA. This result might be explained by the fact that the decrease in the diffusivity at higher pressure leads to a reduction in the interaction between SC–CO_2 and solute contained within matrix. This makes no effective changes in extraction yield of UA at

(A) USC–CO_2 extraction

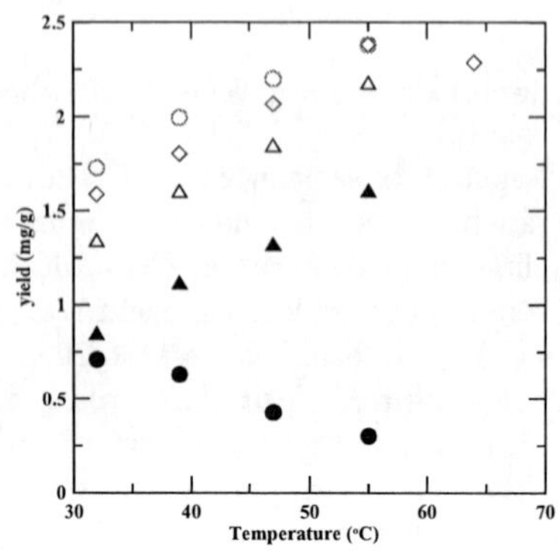

(B) SC–CO_2 extraction

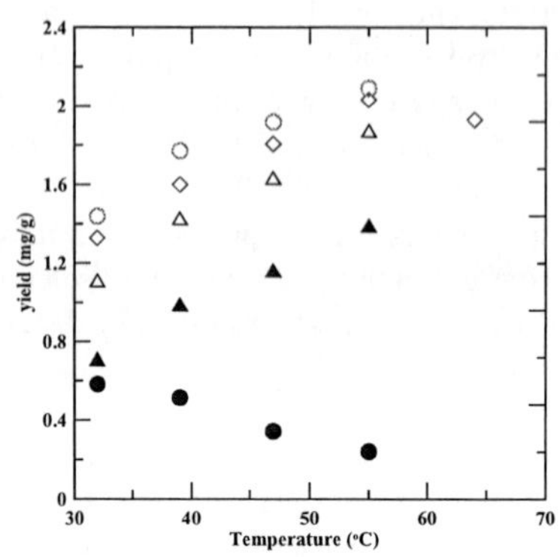

Fig. 2. Effects of extraction pressure and temperature on the extraction yield of OA using (A) USC–CO_2 and (B) conventional SC–CO_2 extractions.
(●:11.0 MPa, ▲: 18.5 MPa, △: 23.5 MPa, ◇:28.5 MPa, ○: 32.0 MPa)

[1]UA yield (mg/g) = weight of the extracted UA/ weight of feeding material.

[2]USC–CO_2 extraction was processed under the static extraction time of 15 min, dynamic extraction time of 80 min, mean plant particle size of 0.36 mm, CO_2 flow rate of 2.1 mL/min with 11.6% aqueous ethanol (80% ethanol) as cosolvent and an ultrasonic frequency of 40 kHz, a power of 185 W and an ultrasound cycle of 75% (intermittent sonication).

[3]SC–CO_2 extraction was processed under the static extraction time of 30 min, dynamic extraction time of 180 min, mean plant particle size of 0.36 mm, CO_2 flow rate of 2.4 mL/min with 11.6% aqueous ethanol (80% ethanol) as cosolvent.

28.5–32.0 MPa, which implied an optimum pressure for practical purposes. The best extraction pressure for the USC–CO_2 extraction of UA from *R. rubescens* with aqueous ethanol as cosolvent was accomplished at 28.5 MPa in this study. Topal et al. [16] also found the similar findings that the increasing pressure from 20 to 40 MPa resulted in a gradual increase in lycopene yield, while further increases in pressure from 40 to 50 MPa did not improve the yield of lycopene.

Because the increase in temperature could have either a positive or a negative effect on the yield of interest compounds, the influence of temperature on SC–CO_2 extraction was more difficult to predict than that of pressure. It is well-known that increasing temperature reduces the density of SC–CO_2, which is not favorable to the dissolution of solute. On the other hand, the saturated vapor pressure of solute increases in according with the temperature, which enables solutes to dissolve in SC–CO_2 more easily. Moreover, the higher temperature the more intense heat motion of solute in aqueous solution. This enables solutes to overcome the adsorbing energy of solution and to desorb more rapidly from active sites of sample matrix by SC–CO_2 at higher temperatures. Consequently, the solubility of solute was likely decreased, kept constant, or increased with rising temperatures at constant pressure, which depended on the balance between these three types of influences competed with each other. Therefore, this temperature-dependent behavior is not quite usual in extraction systems employing SC–CO_2 as the solvent.

By looking at the effect of temperature on the yield of UA as plotted in Fig. 2, a retrograde solubility behavior exists in the supercritical state for the interest compound. At lower pressure of 11 MPa, the extraction yield decreased with increasing temperature, but at the other four higher pressures (18.5, 23.5, 28.5 and 32.0 MPa), the inverse trend was observed. Similar phenomena have been documented for various solid solute in SC–CO_2 system. For example, it is in accord with supercritical carbon dioxide extraction of microalgae lipid [17] and extraction of rice bran oil by supercritical carbon dioxide [18]. At lower pressure (11 MPa), SC–CO_2 density can be largely decreased by increasing temperature, the resultant yield decrease is mainly due to the more dominant effect of the decrease in density. In contrast, the extraction yield increases with the increasing temperature (32–55 °C) at higher constant pressure (18.5–32.0 MPa) because the density of SC–CO_2 is less dependent on temperature at higher pressures, when compared with the decrease of density, increase of vapor pressure and desorb energy. However, with further increase of temperature from 55 to 64 °C, the yield of UA starts to decrease with increasing temperature. In fact, the extraction at this temperature (64 °C) seemed to deteriorate UA where the extracted yield slightly went down to 2.290 mg/g dry plant. Similar findings were evidenced by Patinha et al. [5] for the SC–CO_2 extraction of triterpenic acids from *Eucalyptus globulus* bark, and by Yang et al. [9] for the USC–CO_2 extraction of OA and UA from *S. barbata* D. Don.

To evaluate the efficiency of USC–CO_2 extraction process, the conventional SC–CO_2 extraction (a static extraction time of 30 min followed by a dynamic extraction of 150 min) was also performed, which are reported in Fig. 2. The process efficiency of USC–CO_2 and conventional SC–CO_2 extraction methods were quantitatively related to extraction yield. It is apparent that the yield extracted by SC–CO_2 extraction with ultrasound-assisted using less severe conditions was higher than that extracted by SC–CO_2 extraction without ultrasound-assisted. The increase in extraction efficiencies and yields were attributed to the intensification of mass transport due to physical effects on the surface of the particles through disruption of the cell structure of the sample, enhancing the intraparticle diffusion. Another possible mechanism responsible to the extraction efficiencies of USC–CO_2 procedure is agitation because there are no feasible mechanical stirrers to be used for SC–CO_2 procedure.

3.2 Identification and quantification

HPLC analysis was carried out for the identification and quantification of UA in crude extract. UA standard was used for the identification of peaks in extracts extracted using HRE and USC–CO_2. The chromatogram of the standard shows that UA is eluted at retention time of 57.17 min (Fig. 3A). Figs. 3B and C represented the chromatograms of selected extracts that exhibit similar patterns with the predominant peak at the retention time of 57.17 min, belonging to UA. Consequently, the UA yields are expressed as milligrams of UA /g of dried weight plant (mg/g DW).

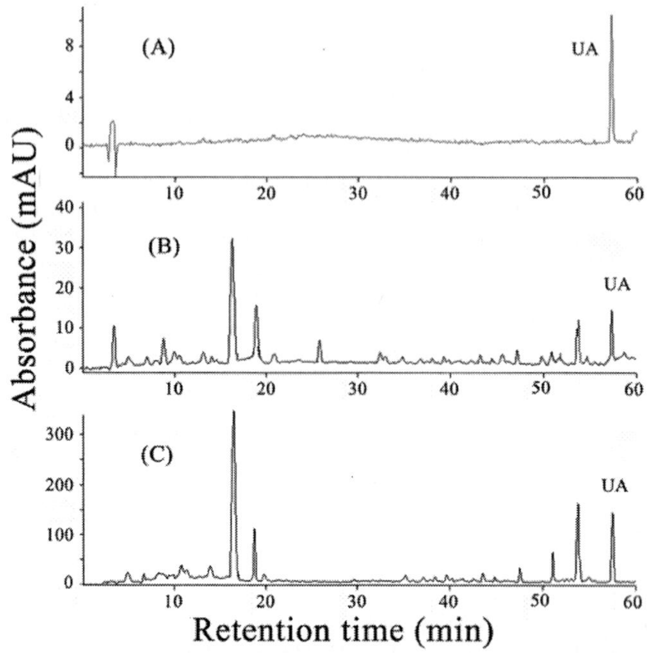

Fig. 3. HPLC chromatograms of (A) the reference compound and *Rabdosia rubescens* extracts obtained using (B) HRE and (C) USC–CO_2 extractions.

3.3 Compared USC–CO_2 extraction and other classical extraction procedures

In order to evaluate the yield of UA in *R. rubescens* crude extract, heat-reflux extraction (HRE) with aqueous ethanol as solvent was performed resulting a maximum UA yield of 1.65 mg on a dry weight basis (mg/g of dry *R. rubescens*). The extraction condition of HRE was shown in Table 1. Compared with the conventional HRE, USC–CO_2 procedure decreases the extraction time, temperature, and solvent consumption with extraction yield of 2.39 mg/g DW. USC–CO_2 procedure shows great potential for efficient large-scale industrial processing and sample preparation for the quality control of UA from *R. rubescens*.

To compare the feasibility of UA extraction from *R. rubescens* using the classical and USC–CO_2 extraction procedure, the extraction yields of UA obtained at both extraction conditions are shown in Table 1. The application of ultrasound during supercritical extraction processes has been proposed as a mechanism both for rate acceleration (115 min vs. 180 min) and yield (2.39 vs. 2.09 mg/g) improvement during comparing with classical SC–CO_2 extraction. Therefore, the study demonstrated that USC–CO_2 extraction shows great potential for efficient large-scale industrial processing and sample preparation for the quality control of UA from *R. rubescens*.

Table 1. Comparison of the extraction conditions and extraction yields obtained using various extraction methods.

Extraction parameters	Extraction mode		
	HRE	SC–CO_2	USC–CO_2
Herbal sample	RR1	RR1	RR1
Mean particle size (mm)	0.36	0.36	0.36
Plant weight (g)	20	20	20
Stirring rate (rpm)	300	—	—
Static extraction time (min)	—	30	15
Dynamic time (min)	—	150	100
Extraction time (min)	60 ×4 (4 cycles)	180	115
Extraction temperature (°C)	Boiling point	55	55
Extraction pressure (MPa)	—	28.5	28.5
Liquid/solid ratio (mL/g)	16	72.0	48.0
CO_2 flow rate (mL/min)	—	2.4	2.1
Extraction cycles	4	—	—
Duty cycle of ultrasound exposure (%)	—	—	75
Percentage of cosolvent (80% ethanol, v/v) in SC–CO_2 (%, v/v)	—	11.6	11.6
UA Yield (mg/g)[a]	1.65 ± 0.06	2.09 ± 0.08	2.39 ± 0.09
Ethanol (%, v/v)[b]	80%	—	—

[a] Values are written as the mean ± SD of six replications and are calculated based on plant dry weight basis (RR1).
[b] Ethanol concentration in water (%, v/v).

4 Conclusions

The results from this study indicate that the extraction of UA from *Rabdosia rubescens* using ultrasound-assisted SC–CO_2 extraction process is feasible. The optimal extraction conditions utilize 28.5 MPa, 55 °C, a 15 min static extraction time, a 100 min dynamic extraction time, a 0.36 mm mean plant particle size, 2.1 mL/min CO_2 and a 40 kHz ultrasonic frequency at 185 W with a 75% cycle (intermittent sonication). Under the optimized conditions, the extraction yield of the UA reached 2.39 mg/g. The results obtained are helpful for full utilization the novel process to extract UA from *R. rubescens*.

References

1. Zhao, J., Deng, J.W., Chen, Y.W., Li, S.P.: Advanced Phytochemical Analysis of Herbal Tea in China. J. Chromatogr. A 1313 (2013) 2–23.
2. Kang N., Cao S.J., Zhou Y., He H., Tashiro S., Onodera S., Qiu F., Ikejima T.: Inhibition of Caspase-9 By Oridonin, A Diterpenoid Isolated from *Rabdosia Rubescens*, Augments Apoptosis in Human Laryngeal Cancer Cells. Int. J. Oncol. 47 (2015) 2045–2056.
3. Wiemann J., Heller L., Csuk R.: Targeting Cancer Cells with Oleanolic and Ursolic Acid Derived Hydroxamates. Bioorg. Med. Chem. Lett. 26 (2016) 907–909.
4. Domingues, R.M.A., de Melo, M.M.R., Oliveira, E.L.G., Neto, C.P., Silvestre, A.J. D., Silva, C.M.: Optimization of the Supercritical Fluid Extraction of Triterpenic Acids from *Eucalyptus globulus* Bark Using Experimental Design. J. Supercrit. Fluids 74 (2013) 105–114.
5. Patinha, D.J.S., Domingues, R.M.A., Villaverde, J.J., Silva, A.M.S., Silva, C.M., Freire, C.S.R., Pascoal Neto, C., Silvestre, A.J.D.: Lipophilic Extractives from the Bark of *Eucalyptus grandis* X *globulus*, A Rich Source of Methyl Morolate: Selective Extraction with Supercritical CO_2. Ind. Crops Prod. 43 (2013) 340–348.
6. Domingues, R.M.A., Sousa, G.D.A., Silva, C.M., Freire, C.S.R., Silvestre, A.J.D., Pascoal Neto, C.: High Value Triterpenic Compounds From the Outer Barks of Several Eucalyptus Species Cultivated in Brazil and in Portugal. Ind. Crops Prod. 33 (2011) 158–164.
7. Yang, Y.C., Wei, M.C., Hong, S.J.: Ultrasound-assisted Extraction and Quantitation of Oil from *Syzygium Aromaticum* Flower Bud (Clove) with Supercritical Carbon Dioxide. J. Chromatogr. A 1323 (2014) 18–27.
8. Wei, M.C., Yang, Y.C.: Kinetic Studies for Ultrasound-Assisted Supercritical Carbon Dioxide Extraction of Triterpenic Acids from Healthy Tea Ingredient *Hedyotis diffusa* and *Hedyotis corymbosa*. Sep. Purif. Technol. 142 (2015) 316–325.
9. Yang, Y.C., Wei, M.C., Hong, S.J. Huang, T.C., Lee, S.Z.: Development/optimization of A Green Procedure With Ultrasound-assisted Improved Supercritical Carbon Dioxide to Produce Extracts Enriched in Oleanolic Acid and Ursolic Acid from *Scutellaria barbata* D. Don. Ind. Crops Prod. 49 (2013) 542–553.
10. Philipe, S., Ana, C. A., Gerardo, F. B., Camila, A.R., Julian, M.: Supercritical Carbon Dioxide Extraction of Capsaicinoids from Malagueta Pepper (*Capsicum frutescens* L.) Assisted by Ultrasound, Ultrason. Sonochem. 22 (2015) 78–88.
11. Yang, Y.C., Wei, M.C.: A Combined Procedure of Ultrasound-Assisted and Supercritical Carbon Dioxide for Extraction and Quantitation Oleanolic and Ursolic Acids from *Hedyotis corymbosa*. Ind. Crops Prod. 79 (2016) 7–17.

12. Wei, M.C., Xiao, J., Yang, Y.C.: Extraction of α-humulene-enriched oil from clove using ultrasound-assisted supercritical carbon dioxide extraction and studies of its fictitious solubility. Food Chemistry, 210 (2016) 172–181.
13. Yang, Y.C., Wei, M.C.: Kinetic and Characterization Studies for Three Bioactive Compounds Extracted from *Rabdosia rubescens* Using Ultrasound. Food Bioprod. Process. 94 (2015) 101–113.
14. Yang, Y.C., Wei, M.C.: Ethanol Solution-Modified Supercritical Carbondioxide Extraction of Triterpenic Acids from *Hedyotis corymbosa* with Ultrasound Assistance and Determination of Their Solubilities. Sep. Purif. Technol. 150 (2015) 204–214.
15. de Melo, M.M.R., Oliveira, E.L.G., Silvestre, A.J.D., Silva, C.M.: Supercritical Fluid Extraction of Triterpenic Acids from *Eucalyptus globulus* Bark. J. Supercrit. Fluids 70 (2012) 137–145.
16. Topal, U., Sasaki, M., Goto, M., Hayakawa, K.: Extraction of Lycopene from Tomato Skin with Supercritical Carbon Dioxide: Effect of Operating Conditions and Solubility Analysis. J. Agric. Food Chem. 54 (2006) 5604–5610.
17. Taher, H., Al-Zuhair, S., Al-Marzouqi, A.H., Haik, Y., Farid, M., Tariq, S.: Supercritical Carbon Dioxide Extraction of Microalgae Lipid: Process Optimization and Laboratory Scale-up. J. Supercrit. Fluids 86 (2014) 57–66.
18. Tomita, K., Machmudah, S., Wahyudiono, Fukuzato, R., Kanda, H., Quitain, A.T.,Sasaki, M., Goto, M.: Extraction of Rice Bran Oil by Supercritical Carbon Dioxide and Solubility Consideration. Sep. Purif. Technol. 125 (2014) 319–325.

Type P63 Digitized Color Images Performs Better Identification for Ovarian Tissue Analysis

T. M. Shahriar Sazzad[1,*], L.J.Armstrong[1], A.K.Tripathy[1, 2]

[1]Edith Cowan University, WA, Australia
{t.sazzad, l.armstrong}c@ecu.edu.au
[2]Don Bosco Institute of Technology, Mumbai, India
amiya@dbit.in

Abstract. *Pathology experts use microscopic biopsy slides for routine examination of different tissues especially ovarian tissues. Substantial amount of processing time is required for this type of manual identification process. Ultrasound is considered as most popular electronic scanning modalities for cancer tissue analysis rather than ovarian tissue analysis. For smaller tissue analysis a more suitable option is to use a computer based approach as it can reduce processing time and effort while increasing the accuracy rate. In this paper a complete review and analysis has been carried out on existing available approaches and a new modified approach has been presented. Comparative results indicate an acceptable accuracy and identification rate.*

Keywords: histopathology; color digitized microscopic image; image artifacts; mean shift; region fusion; cluster; ovarian reproductive tissues.

1 Introduction

According to medical experts, women who do not conceive child before the age of 35 commonly face conceiving complications and expert medical consultation is often necessary. As part of the standard medical consultation process routine examination in the laboratory is required before prescribing any treatment. There are various ways to perform the routine examination process and among them ultrasound is considered as one of the most popular approach. The limitation of ultrasound is that it can only analyze large and more mature tissues and an expert interpretation is also essential for identification [1]. At present, there is no suitable approach which can accurately identify small NGFs (Non-Growing Follicles) using the ultrasound scanning device. Most appropriate approach to date is the manual microscopic identification approach for both large and small NGFs in the histopathology laboratory.

Manual microscopic biopsy slide analysis is considered as the "gold standard" in the laboratory [2]. The limitation of this "gold standard" method is that it is time consuming and has accuracy errors between experts [3]. To overcome the issues associated with microscopic approach computer based analysis approaches are more reliable [3, 4].

Generally tissues are colorless and hard to distinguish from one another [5]. Study of [5] mentioned that dyes or colored organic substances are used by experts in the laboratory during slide preparation due to the fact that dyes or colored chemicals changes tissue colors which assist experts to analyze tissues easily [5]. Different types of microscopic biopsy slides available in the pathology laboratory [6] and type P63 is one of the available slides. Type P63 has two different types which include non-counter stained (tissue stained without eosin or haematoxylin) and counter-stained (tissue stained with eosin or haematoxylin).

Among all available slides Haematoxylin and Eosin (H&E) is considered as most commonly used slide [7]. The main limitation of type H&E is it contains intensity variations [8]. Additionally, type H&E does not perform better identification for small tissues [9, 10]. In compare to H&E, PCNA (Proliferating Cell Nuclear Antigen) performs better identification [7]. Research work by [11] also suggested that type P63 could be a better choice for ovarian tissue analysis.

2 Related Work

Existing research works [1, 7, 12] related to ovarian tissue analysis are mainly semi-automated rather than an automated. Additionally most works are based on animal tissues [6]. At present, there is no research work has been carried out on human ovarian tissues using type H&E [6]. One research work carried out using type PCNA by [11], one research work by [6] using type P63 non-counter stained (100x magnification), one research work by [13] using type P63 counter stained (100x magnification) and a comparative analysis of both counter and non-counter stained was carried out by [14].

An example of type PCNA, type P63 non-counter-stained and counter-stained digitized images of 100x magnification are shown in Fig. 1.

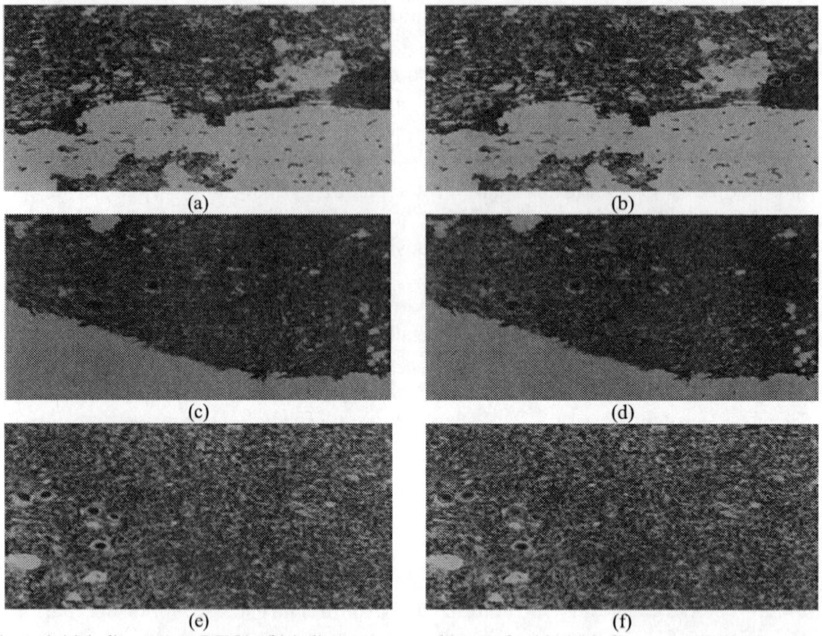

Figure 1.(a) indicates type PCNA, (b) indicates annotated image for (a), (c) indicates type P63 non-counter stained image, (d) indicates annotated image for (b), (e) indicates P63 counter stained image, (f) indicates annotated image of (e). Red marked regions are confirmed nucleus identified by 2 experts and blue marked regions are confirmed by at-least 1 expert.

Currently, all existing research works using type P63 [6, 13, 14] are automated. Proposed work by [11] using type PCNA is not fully automated due to the fact that calibration of processing parameters are essential for a new set of images.

Conventional threshold based approach was implemented by [11] where type PCNA images were used. Type P63 images were used by [6, 13, 14] where image artifact issues were considered. Additionally, study of [13, 14] incorporated some classification approaches as well.

All other existing automated approaches [1, 15, 16] are mainly for cancer cell or tumor cell analysis.

3 Proposed Method

A new modified automated approach has been presented in this study. Additionally, comparison has been carried out for type PCNA, type P63 non-counter and counter stained digitized color images. A detailed flowchart of the proposed method for this study is shown in Fig.2.

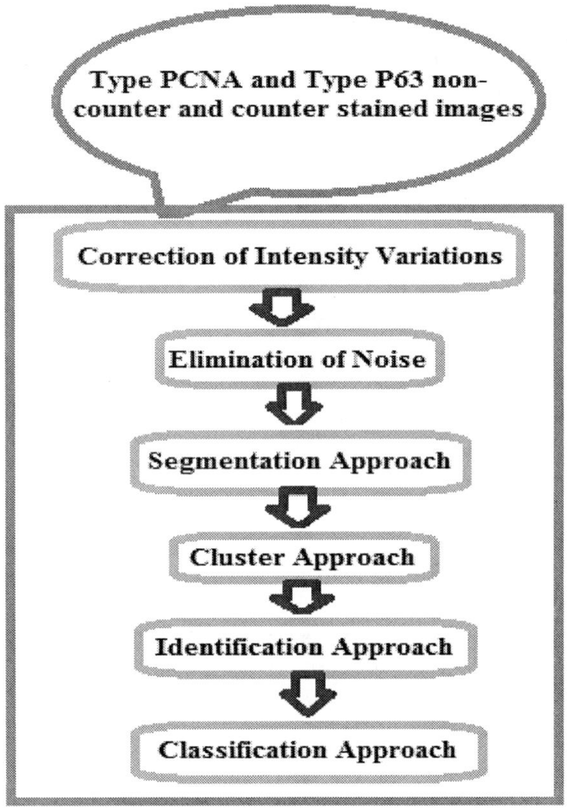

Fig. 2. Block diagram for automated identification process

3.1 Correcting Image Artifacts

Experts use color chemicals in the laboratory during biopsy slide preparation which cause intensity variations [5]. Therefore; it is necessary to correct intensity variations. Research study by [6] mentioned that morphological operation provides more suitable results among all available techniques. Work of [13, 14] also incorporated and proposed modified morphological operation of [6]. Further modification has been carried out in this research study where microscopic eye piece magnification, image magnification and cell diameter were considered to compute the disk radius for morphological operation. This new proposed modified approach was able to provide improved result shown in Fig. 3. Research study test image results are shown in Fig. 4.

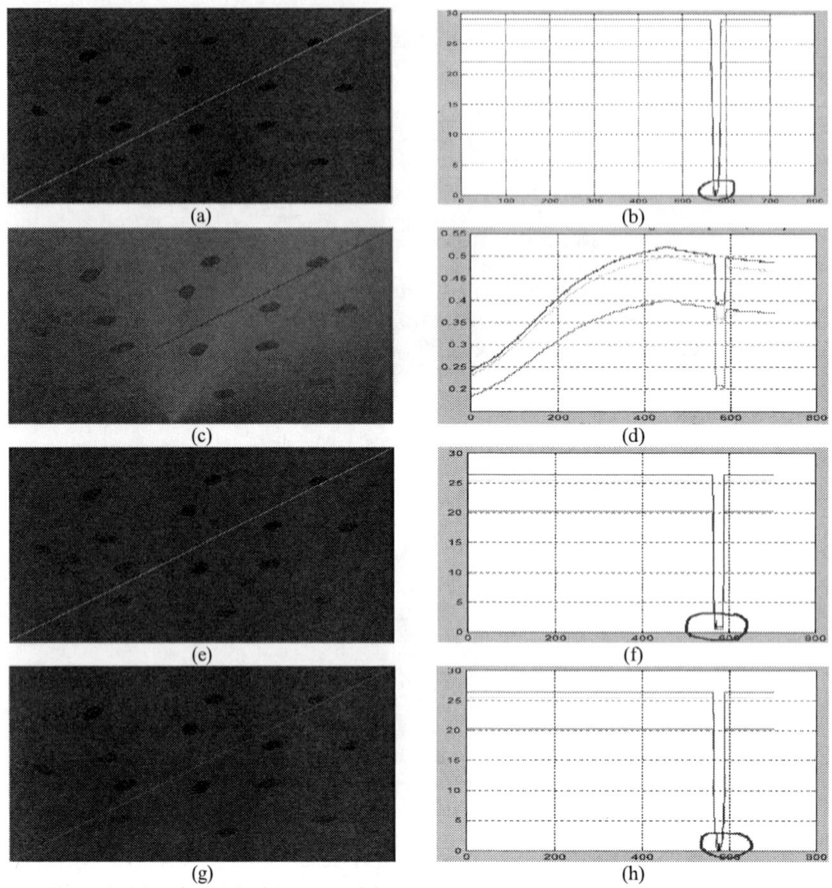

Figure 3. (a) is the original image used from [6], (b) indicates intensity graph across the red line for image (a), (c) is the image with illumination issues and (d) indicates intensity graph across the red line for image (c). (e) indicates corrected image proposed by [6] with intensity graph (f). (g) indicates corrected image of this research study modified proposed and (h) indicates the intensity graph. Intensity graph (f) indicates little dissimilarity in blue circled area but for (h) that area is almost similar to (b).

(a) (b) (c)

Figure 4.(a) indicates type PCNA, (b) indicates P63 non-counter stained and (c) indicates P63 counter-stained corrected images using this study proposed modified approach

3.2 Filter Operation

Among all existing ovarian tissue analysis approaches, median and pixel-based mean-shift filters were used where [6] proposed pixel-based mean-shift filter approach. Study of [13, 14] have incorporated the approach of [6] which was found to provide satisfactory results for type P63. Research work of [11] incorporated median filter which did not provide satisfactory result mentioned by [6]. Therefore; for this research study pixel based mean-shift filter approach was used. The results are shown in Fig. 5.

(a) (b) (c)

Figure 5.(a) indicates type PCNA, (b) indicates P63 non-counter stained and (c) indicates P63 counter-stained filtered images using [6]

3.3 Color Segmentation

For gray-scale image processing threshold and watershed based segmentation approaches are useful but not suitable for color images [11]. The modified region fusion approach proposed by [6] has shown to provide satisfactory results. Modification has been carried out by [13] to improve the results. Further modification of region fusion approach is reported in this research study.

Study of [13] has considered RGB image but in this research study RGB image was divided in 3 different channels (R,G,B). Each channel was indexed to perform the region merging operation. To minimize the standard deviation error this study followed the approach of [6]. Segmentation results are shown in Fig. 6.

(a) (b) (c)

Figure 6.(a) indicates type PCNA, (b) indicates P63 non-counter stained and (c) indicates P63 counter-stained segmented images using this study proposed approach.

3.4 Cluster Approach

Mean shift modified clustering approach proposed by [6] is a suitable approach as it does not require any predefined parameter and shown to provide satisfactory result mentioned by [13, 14]. For this research study clustering approach of [6] was incorporated.

3.5 Identify Ovarian Nucleus

For this research study same features information that were used by [6, 11, 13, 14] for the identification of ovarian nucleus regions were incorporated while eliminating border touching regions [6]. Fig. 7 indicates the identified results for this research study proposed approach.

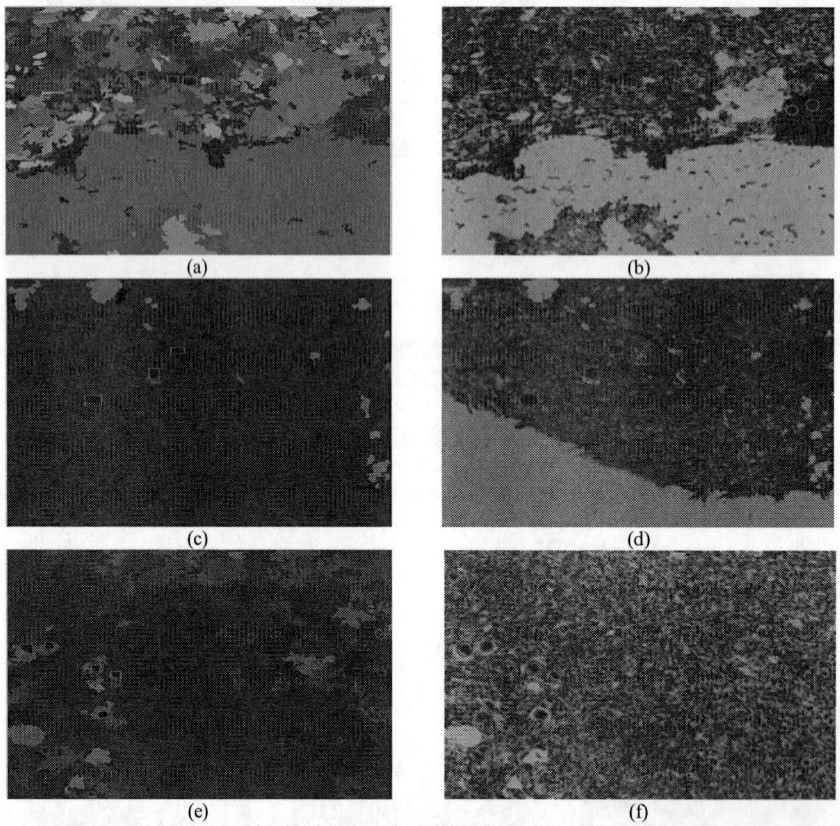

Figure 7. (a) indicates identified regions for PCNA image using this study proposed approach, (b) indicates annotated result from experts. (c) indicates identified regions for P63 non-counter stained images using this study proposed approach, (d) indicates annotated result from experts. (e) indicates identified regions for P63 counter stained images using this study proposed approach, (f) indicates annotated result from experts. One region missed in (a) for type PCNA in comparison with (b). No region missed for type P63 non-counter and counter stained images. No false region identified for any types.

3.6 Classify Identified Regions to Improve Accuracy

Although the indentified results indicates satisfactory results shown in Table II however; a suitable classifier would be a viable option to test the identified regions to improve accuracy. Study of [6, 11] did not apply any classification approach but study by [13] incorporated 3 most popular classifiers (SVM, K-NN and P-NN) from which SVM classifier was most accurate.

Table 1 shows classification accuracy for this research study test images.

TABLE I. CLASSIFICATION ACCURACY TEST USING SVM CLASSIFIER

Name	Stain Type	150 images were used (50 in each group)			Accuracy %	
		Group 1	Group 2	Group 3		
SVM	PCNA	Training	Test	Test	90	90
		Test	Training	Test	89	
		Test	Test	Training	91	
	P63 (non-counter)	Training	Test	Test	95	95
		Test	Training	Test	97	
		Test	Test	Training	93	
	P63 (counter)	Training	Test	Test	96	97
		Test	Training	Test	98	
		Test	Test	Training	97	

4 Experimental Results

Table II indicates that this study proposed modified approach has an improved accuracy over 90% for type PCNA and type P63 which maintains "gold standard" criteria [11].

Study proposed method by [11] do not satisfy "gold standard" criteria as the accuracy rate was found under 80% for all types of images as shown in Table III. Other approaches [6, 13, 14] provides satisfactory results and among all approaches this study proposed approach has the most accuracy rate shown in Table III.

TABLE II. IDENTIFICATION RESULT FOR TYPE PCNA, TYPE P63 (NON-COUNTER AND COUNTER STAINED) IMAGES

Number of test images PCNA (403), P63 non-counter stain (493), P63 counter stain (475)	Image Type	Accuracy Rate (%)
Proposed method	PCNA	90.5
	P63 (non-counter stain)	95.5
	P63 (counter stain)	96.5

TABLE III. COMPARATIVE RESULT FOR ALL AUTOMATED APPROACHES

Number of test images PCNA (403), P63 non-counter stain (493), P63 counter stain (475)	Image Type	avg. processing time (sec)	Precision	Recall
Proposed method	PCNA	24.37	0.91	0.90
	P63 (non-counter stain)	21.76	0.95	0.96
	P63 (counter stain)	22.21	0.96	0.97
Automated approach [13]	PCNA	24.73	0.89	0.90
	P63 (non-counter stain)	22.30	0.95	0.955
	P63 (counter stain)	22.83	0.96	0.965
Automated approach [6]	PCNA	24.55	0.90	0.89
	P63 (non-counter stain)	22.57	0.96	0.95
	P63 (counter stain)	23.02	0.955	0.965
Automated approach [11]	PCNA	26.80	0.79	0.74
	P63 (non-counter stain)	24.10	0.75	0.73
	P63 (counter stain)	24.46	0.76	0.77
Automated approach [14]	PCNA	24.37	0.875	0.91
	P63 (non-counter stain)	22.12	0.945	0.96
	P63 (counter stain)	22.60	0.96	0.96

5 Discussion and Conclusion

Existing automated computerized approaches for ovarian tissue analysis were reviewed in this research study. All automated cancer cells identification approaches were not considered due to the fact that cancer tissues are different from ovarian reproductive tissues.

Research study test image results indicates that type P63 digitized color images performed better identification for ovarian tissue analysis in compare to type PCNA. This is a novel study due to the fact that this is the first published study where type PCNA and type P63 digitized images were compared for ovarian tissue identification. Additionally, a significant amount of type PCNA and type P63 (non-counter and counter) stained images were used for comparison and analysis purposes. If expert observation variability can be minimized then it would be possible to increase the accuracy rate.

6 Acknowledgement and Future Work

The authors would like to thank Assistant Professor and head of the department (Pathology) and domain expert Doctor S.I. Talukder (MBBS, M.Phil (Pathology), Shaheed Sayed Nazrul Islam Medical College, Kishoreganj, Bangladesh) for providing the test images, annotated images and necessary feature information. In future this research study will review other stains if the images can be collected from the experts.

References

[1] A. Skodras, S. Giannarou, M. Fenwick, S. Franks, J. Stark, and K. Hardy, "Object recognition in the ovary: quantification of oocytes from microscopic images," in *Digital Signal Processing, 2009 16th International Conference on*, 2009, pp. 1-6.

[2] V. Kiruthika and M. Ramya, "Automatic Segmentation of Ovarian Follicle Using K-Means Clustering," in *Signal and Image Processing (ICSIP), 2014 Fifth International Conference on*, 2014, pp. 137-141.

[3] L. Muskhelishvili, S. K. Wingard, and J. R. Latendresse, "Proliferating cell nuclear antigen—a marker for ovarian follicle counts," *Toxicologic pathology,* vol. 33, pp. 365-368, 2005.

[4] M. R. Lamprecht, D. M. Sabatini, and A. E. Carpenter, "CellProfiler™: free, versatile software for automated biological image analysis," *Biotechniques,* vol. 42, p. 71, 2007.

[5] D. Magee, D. Treanor, D. Crellin, M. Shires, K. Smith, K. Mohee, *et al.*, "Colour normalisation in digital histopathology images," 2009.

[6] T. Sazzad, L. Armstrong, and A. Tripathy, "An Automated Detection Process to Detect Ovarian Tissues Using Type P63 Digitized Color Images," in *Tools with Artificial Intelligence (ICTAI), 2015 IEEE 27th International Conference on*, 2015, pp. 278-285.

[7] C. A. Picut, C. L. Swanson, K. L. Scully, V. C. Roseman, R. F. Parker, and A. K. Remick, "Ovarian follicle counts using proliferating cell nuclear antigen (PCNA) and semi-automated image analysis in rats," *Toxicologic pathology,* vol. 36, pp. 674-679, 2008.

[8] T. Mouroutis, S. J. Roberts, and A. A. Bharath, "Robust cell nuclei segmentation using statistical modelling," *Bioimaging,* vol. 6, pp. 79-91, 1998.

[9] T. J. Bucci, B. Bolon, A. R. Warbritton, J. J. Chen, and J. J. Heindel, "Influence of sampling on the reproducibility of ovarian follicle counts in mouse toxicity studies," *Reproductive Toxicology,* vol. 11, pp. 689-696, 1997.

[10] B. Bolon, T. J. Bucci, A. R. Warbritton, J. J. Chen, D. R. Mattison, and J. J. Heindel, "Differential follicle counts as a screen for chemically induced ovarian toxicity in mice: results from continuous breeding bioassays," *Toxicological Sciences,* vol. 39, pp. 1-10, 1997.

[11] T. W. Kelsey, B. Caserta, L. Castillo, W. H. B. Wallace, and F. C. Gonzálvez, "Proliferating cell nuclear antigen (PCNA) allows the automatic identification of follicles in microscopic images of human ovarian tissue," *arXiv preprint arXiv:1008.3798,* 2010.

[12] P. Soucek and I. Gut, "Cytochromes P-450 in rats: structures, functions, properties and relevant human forms," *Xenobiotica,* vol. 22, pp. 83-103, 1992.

[13] T. Sazzad, L. Armstrong, and A. Tripathy, "An Automated Approach to Detect Human Ovarian Tissues Using Type P63 Counter stained Histopathology Digitized Color Images " in *IEEE-EMBS International Conference on Biomedical and Health Informatics (BHI)*, 2016, pp. 25 - 28.

[14] T. Sazzad, L. Armstrong, and A. Tripathy, "A Comparative Study of Computerized Approaches for Type P63 Ovarian Tissues Using Histopathology Digitized Color Images," in *10th International Conference on Computer Graphics, Visualization, Computer Vision and Image Processing (CGVCVIP 2016)*, Portugal, 2016.

[15] G. Landini and I. Othman, "Estimation of tissue layer level by sequential morphological reconstruction," *Journal of microscopy,* vol. 209, pp. 118-125, 2003.

[16] M. Yoshida, A. Sanbuissyo, S. Hisada, M. Takahashi, Y. Ohno, and A. Nishikawa, "Morphological characterization of the ovary under normal cycling in rats and its viewpoints of ovarian toxicity detection," *The Journal of toxicological sciences,* vol. 34, pp. SP189-SP197, 2009.

Author Index

Journals by ibai-publishing

The journals are free on-line journals but having in parallel hardcopies of the journals. The free on-line access to the content of the paper should ensure fast and easy access to new research developments for researchers all over the world. The hardcopy of the journal can be purchased by individuals, companies, and libraries.

Transactions on Machine Learning and Data Mining
(ISSN: 1865-6781)

The International Journal "Transactions on Machine Learning and Data Mining" is a periodical appearing twice a year. The journal focuses on novel theoretical work for particular topics in Data Mining and applications on Data Mining.

Net Price (per issue): EURO 100
Germany (per issue): EURO 107 (incl. 7% VAT)

Submission for the journal should be send to:
info@ibai-publishing.org

For more information visited: www.ibai-publishing.org/journal/mldm/about.html

Transactions on Case-Based Reasoning
(ISSN:1867-366X)

The International Journal "Transactions on Case-Based Reasoning" is a periodical appearing once a year.

Net Price (per issue): EURO 100
Germany (per issue): EURO 107 (incl. 7% VAT)

Submission for the journal should be send to:
info@ibai-publishing.org

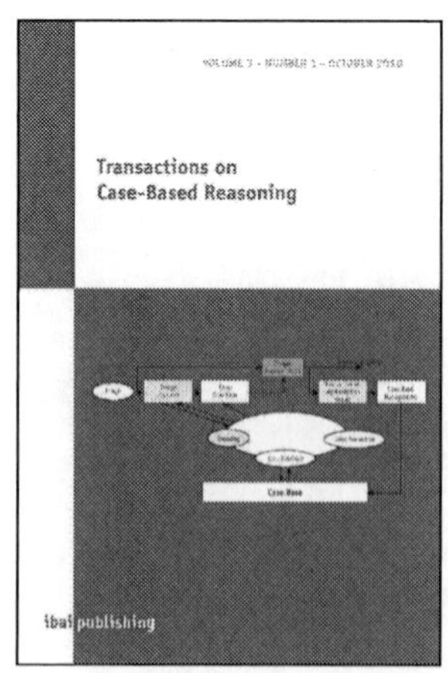

For more information visited: www.ibai-publishing.org/journal/cbr/about.html